Voices From Vietnam:
Stories of War

by

Kayleen Reusser

Kayleen Reusser Media

Voices From Vietnam: Stories of War first published in the United States by Kayleen Reusser Media. Printed in the United States.

Copyright © 2022 by Kayleen Reusser

All rights reserved. No part of this publication may be reproduced or distributed in any form or by any means, or stored in a data base or retrieval system, without the prior written permission of the author.

KayleenReusser.com

ISBN: 978-1-7325172-9-5

Book Cover design by Rob Williams of www.ilovemycover.com.

Printed in the United States of America

The information provided within this book is for general informational purposes only. While the author has tried to provide correct information, there are no representations or warranties, express or implied, about the completeness, accuracy, reliability, suitability or availability with respect to the information, products, services, or related graphics contained in this book for any purpose. Any use of this information is at one's own risk.

Photographs are courtesy of individuals as named and the author.

Contents

Michael Blough -- Army 11

Michael Boles -- Army 17

Alfred Brothers – Air Force 25

Fritz Bultemeyer -- Army........................31

Mike Chamness -- Army........................ 43

Dennis Covert -- Navy............................ 52

Richard Dawson -- Marines..................... 57

Mike Dean -- Army 64

Randy Harnish – Army 69

Arley Higginbotham – Army 81

Jose Huerta – Navy 90

Lanny Idle – Air Force 99

LeRoy Jesfield -- Army........................106

Dan Lavine – Marines 112

Rod Maller -- Army 117

Tom Paxson – Army 127

John Senac – Army 132

Harold Stanford -- Army 140

Book Club Discussion Questions 152

Author's Note:

This book is in answer to the many requests I have received over the years since interviewing veterans. While my focus of writing books and speaking has been about World War II, at nearly every venue someone would say, "When are you going to write a book about Vietnam?"

Upon choosing to devote my efforts to a book on stories of veterans who served in this war, I found several items of interest, including the overwhelming presence of politics during the war, division of Americans with war protesters objecting to the soldiers being in Vietnam, and the ability to get out of military service by attending college.

I was too young to be part of any of that, but maybe that's good as I can come to this project with an open mind. Strictly speaking, that's not true, as the interests of the American military is where I'll be standing when the dust clears.

So in tribute to the 2,700,000 men and women who served and possibly didn't come home to a hero's welcome, I say, "Welcome Home."

Acknowledgements

Thanks to Mark Miller, an Army veteran, for publishing my military stories for many years in the News-Banner, my hometown newspaper.

Thanks to Glenda Nichols-Oliver and Bonny Heisterkamp of Gemini Concepts, Inc. for also being strong supporters of my military writing. Much appreciated!

Thanks to all of the veterans I've interviewed over the years. Your stories keep me inspired to keep working.

Thanks to my family, especially my Vietnam War-era husband who is retired from the Air Force. He has done everything to support my efforts from introducing me to veterans to taking me on tours of Europe to paying for the gas to interview veterans to hooking up equipment for my talks and driving me home afterward. He also cooks, cleans, shops, and does laundry.

And no, he is not for sale or rent.

God bless America and those who keep Her safe!

Foreword

by Sammy L. Davis

From 1965 through 1967, I was at war.

Like my grandfathers and my father, I had volunteered to serve as a United States Army soldier. At first, I was doing what our family did – serving America in the military in its time of need.

The main difference was that I was sent to serve in a country none of my ancestors had stepped foot in – Vietnam.

Many Americans had no idea what my fellow soldiers and I were doing halfway around the world. We were on another continent fighting against people who seemed to have little that compared to life in America.

Many Americans must have asked -- why should our country care about them?

At first, we soldiers knew little more than our families about the challenges the Vietnamese people faced and the difficulties they endured since war had arrived on their doorsteps.

Starting the first day we put boots on the ground, each of us troops received an eye-opening education from 100-degree heat to the food and living conditions among the Vietnamese people to fighting in triple-canopy jungles against people who seemed to want to eradicate freedom from the South Vietnamese people. Along the way, we gained -- and lost -- forever comrades and friends.

Sadly, coming home at the end of our tours should have been a relief. Instead, we faced another divided nation that blatantly turned its back on us. That was tough for all of us.

Now, more than 50 years later, memories of war remain with many. Some have leaned on religious faith to get through the long nights. Others still struggle with questions about the past in which we were encouraged to give something greater than ourselves.

As a result of receiving the nation's highest military honor for valor in combat, the Congressional Medal of Honor, I have had the great pleasure to get to know many veterans over the years through speaking and attending events.

I feel an affinity with each soldier and those whose stories are told in this book. They fought for their country in Vietnam, like I did, with one hundred percent effort and pride.

I have always believed each veteran, no matter his or her era of service, deserves respect, dignity, and a lifelong recognition of commitment to duty. In my humble opinion, this seems especially true of this group that left as young men and returned much more mature in their outlooks about life.

Now it is America's turn to serve them by listening to what they have to say. Let us never forget about the price of freedom and the efforts of those who gave their all.

Welcome Home, my brothers in arms.

    Sammy L. Davis, Medal of Honor recipient, Vietnam

Michael Blough – Army

It was a high-risk mission.

Michael Blough's flight crew had been ordered to rescue injured soldiers who had been on patrol in the jungles of Vietnam. The helicopter pilot would enter the landing zone and hover while Blough and the others dismounted and loaded wounded soldiers, some of whom appeared to have lost limbs.

Blough knew the task was dangerous, though he didn't know how dangerous until the pilot told him he was in a mine field. When the pilot called him a hero, Blough, 19, shook his head. "I had not known the severity of the situation," he insisted. "I was just doing what I had been trained to do."

**

Blough, a native of South Bend, Indiana, enlisted in the Army in 1969 for personal reasons. "I wanted to be somebody special," he said. "I thought being an American soldier would give me status." Blough had a good example of a courageous soldier -- his father had served at Pearl Harbor. The younger Blough hoped to emulate his father's patriotic service.

In Vietnam Michael Blough served as a door gunner on helicopters tasked with firing and maintaining manually-directed armament. He participated in dozens of missions, sometimes under enemy fire.

Performing dangerous acts in a country at war created a tense environment. Housed together in barracks, the American soldiers sometimes shared friendships in which it was comfortable to express thoughts of fear of being captured by the North Vietnamese. Others shared fears about dying.

One day, a pilot named Paul Nome commented that he felt he would die in Vietnam. Nome was a pilot whose skills Blough respected as they flew together on missions.

As a teen, Mike Blough had survived a house fire. Having already looked death in the eye, coupled with the fact that he attended church and always had faith God would save him from troubles, he was not afraid.

A few days later, Blough, who had been promoted to crew chief, was given a mission: he and his crew would fly in a helicopter over an area filled with foliage to look for enemy soldiers. The pilot would fly low to detect body heat, which could mean hitting a tree, stalling, or crashing. "Our crew knew we might not come back," said Blough.

On the morning of the mission, Blough walked along the flight line, looking for his chopper. It was gone. He was told Nome had left earlier with the rotary-winged aircraft to fulfill the risky assignment.

Blough tried not to worry, as he recalled his friend's grim prediction. Later that day, he received the devastating news that a rocket-propelled grenade had crashed through the chopper's cockpit and detonated.

Nome survived the attack and Blough raced to the hospital to be with his friend. He was shocked to see Nome's lower face heavily wrapped in bandages because his chin had been blown off. Blough could tell his friend was on his way out.

As Nome's eyes closed, Blough was devastated, yet glad in knowing his friend would never feel pain or uncertainty again.

Blough felt the loss of his friend's life keenly. He believed the only way he could redeem his friend's death was to re-enlist.

In 1970, he was assigned as crew chief to the 129th Attack Assault Helicopter Company flying "Hueys" and Cobra helicopters.

When his company was sent to Germany, they became the 124th Cavalry. "We monitored radio communications," he said.

In July 1976, Blough was discharged at the rank of E5/Spec 5. He married and became the father of five children. Blough taught his offspring about patriotism. "This country was founded on principles that I want to see upheld for the next generation," he said. "We have talked about what it means to be an American and listened to Red Skelton's recitation of the Pledge of Allegiance."

For two decades Blough worked for the Indiana Department of Corrections. In 2015, he resigned, following a diagnosis of reflex sympathetic dystrophy in his hands. His symptoms, which included complex and regional pain syndrome involving extensive inflammation, were thought to be the result of exposure to Agent Orange in Vietnam.

Since his return home from Vietnam, Blough's mind suppressed his time there, including his experiences with Nome. He only recalled the incidents following an unexpected development during surgery in May 2013 when his heart stopped.

He was successfully resuscitated on the operating table, somewhat to the amazement of the medical staff. Then the memories flooded over him. It took a while for Blough to navigate through them.

When Blough shared his times in Vietnam with family, a daughter said she believed he would have helped the wounded soldiers in the minefield, whether or not he had known it was dangerous. Blough remained unconvinced. "I may have gone in to rescue those guys, but I would have liked to have known it was a minefield," he said. "Then I would have tiptoed."

Today, Blough believes he was revived for a reason. "Many people I knew from 'Nam were messed up after they went home," he said. "I think I was not screwed up because I suppressed most of it. But my friends at Cornerstone Community Church in Decatur, Indiana, have helped me stay tuned to God. He's still taking care of me today."

Despite the loss of friends and health problems, Blough refuses to be bitter. "When you're young, you think you won't die and that you'll live forever. I learned death can be at your door any time, any minute. I owe God a lot for keeping me safe."

E5/Spec 5 Mike Blough

Michael Boles – Army

The earth smelled dank and too close. Michael Boles forced himself to breathe slowly as his slight frame was lowered head first into the small opening in the ground. Soldiers from his platoon held tightly to his legs as he entered, inch-by-inch, the dark, unknown interior.

The Army called this and hundreds of other openings found throughout the Vietnamese countryside, "spider holes." The Viet Cong had dug them to hide, sometimes adding tunnels jutting to the sides. Some holes discovered by squadrons led to intricate labyrinths of rooms beneath buildings.

All held the potential of danger.

At 130 pounds and one of the smallest soldiers in his company (he had lost 30 pounds since arriving in the country), Boles was designated the "tunnel rat" in search of the holes made by the enemy.

Carrying a .45 pistol and flashlight while on the lookout for punji sticks, snakes, and C-ration cans with sharp edges, Boles had investigated dozens of spider holes.

Once, when a hole appeared to be deeper than normal, Boles went in feet first. Shining his flashlight below him, at first he saw nothing.

Then his light landed on a foot.

Shooting his weapon in the hole, Boles yelled to be hauled up, which his unit did quickly, tossing grenades into the hole afterward. Later, they pulled lifeless bodies of enemy soldiers from the rubble.

**

Michael Boles was born in Fort Wayne, Indiana, in 1947. After graduating from Northside High School in 1965, he worked at General Electric for a year to earn money for college. In October 1966, two days after beginning classes at International Business College (he wanted to be an accountant), his draft notice arrived.

When Boles applied to the draft board for a deferral from military service due to his enrollment in college, the board said no. They reasoned his paperwork had been processed before he started classes.

After completing basic training at Fort Leonard Wood in Missouri, Boles trained as an infantry soldier at Fort Polk, Louisiana. The Army nicknamed the area "Little 'Nam" for its hot weather, humidity, and plethora of bugs. Boles was assigned to "Infantry 11 Bravo" unit.

It seemed an anomaly to Boles that he, a city boy, scored the highest for firing an M16, especially since he wore glasses.

Eyewear would be a problem in Vietnam as Boles often misplaced his glasses. He was thankful to his family for responding promptly to letters for help in replacing them. Even though it usually took them a month to get a pair from their eye doctor, Boles believed that was quicker than waiting for the military to issue them.

On April 1, 1967, Boles and other recruits filled a TWA flight from Fort Wayne, landing at Bien Hoa Army Base, a military airfield in south-central Vietnam, 16 miles from Saigon.

Boles was re-assigned to the 196th Infantry. His company was sent to a missile base at Da Nang, 50 miles south of Bien Hoa where his company was assigned to Hill 69 and Hill 79.

For three months, Boles' duties consisted of mostly guard patrols. It was mostly mundane, although sniper fire occurred often at night.

Later, Boles accompanied his squadron on search-and-destroy missions. Trudging through water-filled streams and paths during the monsoon season, many soldiers developed foot fungus. The military tried to alleviate this problem by re-supplying the soldiers with clean uniforms and socks every fourth day when helicopter pilots dropped parcels over their location.

A soldier exits a hole in the ground after searching for signs of the enemy. Courtesy Rod Maller.

Soldiers on patrols were also issued two cans of pop, two cans of beer, and a crate of C-rations to last them four days. No matter how hungry he got, one item remained in Boles' C ration crate unopened, that of ham and Lima beans. "No one ate them," he said.

He ate so much peanut butter that upon returning home he never touched it again.

Even the water tasted foul. At Boles' request his family sent packages of dry, sweetened Kool-Aid. The support of loved ones with supplies and letters meant much to Boles. "For us soldiers, mail from family and friends was our lifeline," he said. Boles wrote back as often as he could.

Sometimes soldiers were tempted by food offered by locals. When two South Vietnamese youth offered him cooked hamburger patties, Boles purchased them. Later, he spied the teens grilling pieces of meat which he suspected had come from a dog. He regretted his earlier purchase.

In March 1968, Boles contracted malaria. The Army medic tried to treat it by giving him quinine pills. When Boles' stomach couldn't tolerate the medication, he stopped taking them and eventually recovered from the malady on his own.

On another occasion, a grenade exploded near Boles during a battle, leaving him injured, though not severely. He received a Purple Heart.

Boles' "Zero Day" (the day he was due to leave Vietnam) was set for April 1, 1968. When the Army issued orders for him to go four days early, it felt to him like being let out of school.

Mike Boles returned to the United States on a commercial flight carrying a North Vietnamese rifle he had found it in a spider hole. Before leaving the country, he had obtained the necessary paperwork.

On the flights to Seattle and then to Chicago, Boles placed the weapon on the seat beside him in the plane's cabin. When he tried to do the same for the short flight from Chicago to South Bend (he would then catch another flight to Fort Wayne), a female stewardess asked if he would consent to allowing his rifle to stay with the pilots in the cockpit. Some passengers had expressed concern about its presence in the cabin. Boles agreed.

He arrived in the United States on March 27, 1968, and was discharged on October 12, 1968 at the rank of Spec 4. He was awarded a Bronze Star for meritorious service.

Boles returned to General Electric where he stayed for 38 years as a machinist. He and his wife, Kathy Hammond, whom he married in August 1968 at Fort Hood, Texas, became parents to three children.

In recent years, Mike Boles has struggled with complications from diabetes and other health concerns, which the medical community believes could be attributed to exposure to Agent Orange.

Yet, Boles does not regret his time in the military. One particular member of his family set the bar high for him in patriotic service.

Gladys Young, his maternal grandmother, donated 20,000 hours over several decades at the Veteran's Administration Hospital in Fort Wayne, organizing Christmas parties, purchasing gifts for veterans, and bringing in cookies.

"Being a soldier was one of the biggest things in my life," said Mike Boles. "By serving my country as a soldier, I know I made my grandmother and other members of my family proud."

Spec 4 Mike Boles

Lt. Colonel Alfred Brothers, Jr.
Air Force

In 1970, Captain Alfred 'Al' Brothers flew his B-52 Stratofortress over northern Laos on a routine mission with his squadron searching for enemy movements. Brothers and other American pilots were stacked in formation, 500 feet apart, separated by one-and-a-half miles. They remained radio-silent from take-off to exiting the combat area because the North Vietnamese monitored frequencies

On each of his combat missions over Vietnam from Guam, Okinawa, and Thailand (before completing his tour of duty, he would perform 68), Brothers looked for trucks being driven by the North Vietnamese. The vehicles were thought to be filled with war materials and soldiers. His usual count of trucks eliminated per mission was two.

By 1970, Brothers had trained as an aviator for several years in a plethora of aircraft. That day, in the skies over northern Laos, his hundreds of hours of experience, combined with an inborn skill and innate instinct, helped him eliminate 10 trucks.

The mission's dangerous location and Brothers' dedication prompted him to be recommended for the Air Force's highest rating – Distinguished Flying Cross.

Since 1926, this medal has been awarded to people displaying an act of heroism or extraordinary achievement while participating in aerial flight during an operation not considered routine.

Brothers did not consider himself a hero or superhuman. He and every pilot knew the truth: each flight could be his last.

The cause could be malfunction, weather, or southeast Asia's treacherous topography, which is mainly a mixture of limestone, dolomite, and gypsum called karst. It created a non-reflective surface that American radar often failed to pick up, causing pilots to crash into mountains that suddenly loomed in front of them.

More likely, however, the cause for a plane or helicopter crash was due to antiaircraft fire by the North Vietnamese. It pierced the sides of a plane and interfered with the engine.

**

Born in 1942 in Boston, Brothers had become interested in aviation at a young age after joining an Air Explorer Post in Boy Scouts. There, he met pilots who inspired him to think of aviation as a career.

At Boston University Brothers joined the Air Force Reserve Officers' Training Corps (ROTC) and its flight instruction program. After graduating in 1964 with a degree in aeronautical engineering, he enlisted in the Air Force and learned to fly B-52s, B-57s, and FB-111s.

In 1964, Brothers spent a year at Williams Air Force Base in Arizona training in T-37s and T-38s. At Minot Air Force Base in North Dakota, he was made part of an accelerated pilot program with B-52s, making him, at age 27, one of the youngest aircraft commanders.

After being sent to Thailand, Brothers finished his tour and returned to the United States to undergo B-57 training at Hill Air Force Base in Utah and at MacDill Air Force Base in Florida.

In 1971, Brothers arrived at Ubon Royal Thai Air Base where he was assigned a squadron of B-57s. The base, approximately 37 miles east of the Laos border near the city of Ubon Ratchathani, was the largest front-line facility of the United States Air Force in Thailand from 1962 through 1975.

The B-57s were equipped with canisters of napalm (an incendiary mixture of a gelling agent and a volatile petrochemical, such as gasoline), fire-bombs, and four laser-guided 500-pound "smart" bombs. To stay above the 37-millimeter ground fire and hills in Laos, Brothers' crew flew 380 miles per hour at 10,000 feet.

At the end of his second tour in Southeast Asia in June 1972, Brothers returned to the United States to re-enlist. He was assigned to Pease Air Force Base in New Hampshire, flying FB-111s. He also flew at Plattsburgh Air Force Base in New York, which served as the FB-111 training base.

Using the plane's automatic terrain avoidance radar system and onboard radar navigational systems, the crew navigated low-level flying in all types of weather without t global positioning system (GPS), even at night. Brothers qualified as an instructor for night (low-level) flying. Brothers also flew as part of Red Flag, the Air Force's premier air combat training exercise at Nellis Air Force Base in Las Vegas.

In 1976, Brothers worked as an aeronautical engineer at Wright-Patterson Air Force Base in Dayton in the Foreign Technology Division. Originally he worked in the Space Systems Division evaluating satellites and reconnaissance systems. He also served as branch chief for Missile Facilities Branch.

In the 1980s Brothers became commander of the Air Force ROTC detachment at Wright State University. It was gratifying for him to return to the program that had established his desire for an aviation career. During his tenure, the ROTC detachment at Wright-Patterson grew to be one of the largest in Ohio.

In 1986, after 22 years of active military service, Brothers retired at the rank of Lieutenant Colonel. Today, he lives in northern Indiana with his wife, Sandi. He stays busy with his family and his involvement in Rotary and Boy Scouts.

"My Vietnam experience provided me the opportunity to engage in wing and squadron operations," he said. "The experience taught me about flexibility and adaptability, both in peacetime and war time."

**

According to the National Archives, approximately 2,545 pilots were killed on missions during the Vietnam War.

Lt. Colonel Alfred Brothers, Jr.

Fritz Bultemeyer -- Army

"Enemy troops coming over the walls! They're throwing grenades! I need to get to my defense station!"

On February 1, 1968, Sergeant Fritz Bultemeyer was speaking with the single-side band (SSB) site operator within the compound at Long Binh, located between Bien Hoa and Saigon. When the operator's communique was cut off, Bultemeyer thought he knew why.

Two days earlier, 85,000 North Vietnamese forces had launched an aggressive assault against dozens of South Vietnamese sites. These included Saigon, the air base of Ton Son Nhut, and military base at Long Binh. The site operator at Long Binh had attempted to relay information to Bultemeyer before being cut off.

Bultemeyer couldn't know if the insurgents had cut power lines or worse, killed the operator.

Upon learning of the assault, which occurred on the Lunar New Year, a holiday called Tet, Bultemeyer contacted SSB stations within his area to discover their status. He knew communications within the system could be cut off at any moment. As the non-commissioned officer in charge of the single-side band, Bultemeyer had hoped to forestall destruction of the military's communications infrastructure.

When he tried re-connecting with the SSB site operator at Long Binh, Bultemeyer had no success. Thirty minutes later, he was relieved to hear the operator's voice back on the air, relaying that its Quick Reaction Forces (QRF) had arrived to drive the enemy back just as the base's walls were being breached.

Sadly, several QRF rescuers had died, though none of the SSB operators. Bultemeyer, who had been raised attending a Missouri Synod Lutheran Church, sent up prayers for those killed in action.

Now, his own compound was surrounded, a situation that would continue for six days. He wondered if his own death would soon follow.

**

Fritz Bultemeyer had been drafted into the Army in 1966. He left his hometown of Decatur, Indiana, to complete basic training at Fort Leonard Wood in Missouri. In February 1967, he flew with other recruits from Travis Air Force Base in California to Saigon. There, Bultemeyer was assigned to the Army Signal Corps as the non-commissioned officer in charge (NCOIC).

Bultemeyer's compound was the control station for the single-side band system and he served as the "net control station operator." As the primary intelligence system supervising radio sites throughout the country, the SSB Net was involved in thwarting major Viet Cong and NVA operations.

All stations in Bultemeyer's net had to go through his station to transmit to him or other stations in the network. His usual activities included monitoring traffic, maintaining message logs, keeping a chain of command on the compound, and informing those in command of what was happening throughout the country.

As the enemy surrounded Bultemeyer's compound, he installed safeguards to ensure no unnecessary or unauthorized traffic was transmitted through the SSB. Should the station's defensives be breached, he would destroy sensitive documents, files, and equipment in the security vault with thermite explosive devices.

Bultemeyer could hear a squadron of the 11th Armored Cavalry Regiment and other Americans fighting to prevent the offenders from taking over the compound. He gritted his teeth in frustration as he recalled how in the fall of 1967, a South Vietnamese sergeant had reported signs of a buildup of North Vietnamese forces in the south.

The sergeant named Nguyen had alerted Bultemeyer to his observations that an increased number of funerals were occurring around Saigon. In Vietnamese culture, funerals were major celebrations of the departed going to be with their ancestors. The sergeant feared the revered events were being used to cover an increase of NVA moving to the area as bereaved family members.

When Nguyen had relayed the information to his chain of command, his observations had been dismissed. He asked Bultemeyer to report it, only for him to receive a similar reaction.

Post-attack evaluations would show fake caskets filled with weapons, armaments, and explosives had been buried in shallow graves by enemy troops for the upcoming assault. In addition, an increased number of fireworks were set off in apparent celebration of the Vietnamese New Year, their raucous cacophony providing an excellent means of disguising the noise of small arms fire.

Bultemeyer felt some relief in recalling that in October 1967 the American intelligence community had beefed up its communication capability by switching their administrative net to the SSB.

A soldier eats C-rations while on patrol. Courtesy Rod Maller.

Bultemeyer and an SSB team had traveled to key collection places to put the administrative net equipment into place and trained personnel on its operation. As a powerful FM system with components that took up little space, the voice-only system left secure lines for critical intelligence situation reports.

Bultemeyer began installing back-up power rectifiers that could boost the local 40Hz electrical current from the South Vietnamese power grids to the station's required 60Hz electrical current. When the station's power-generating equipment (which produced 60Hz electricity for the station's secure radio, teletype, encrypting, and SSB equipment) were tampered with and destroyed, Bultemeyer tried to cope with them, too.

He also tried to squeeze in training sessions with personnel to take over his duties, should he be unable.

As it turned out, the NVA failed to interrupt communications and intelligence capabilities as only a few phone lines were disrupted and no 40Hz power lines to the compounds were severed.

At 2200 hours, Bultemeyer received reports that the SSB site at Da Nang was surrounded and taking small-arms fire. Not only that, but their generator had been hit by shrapnel from a mortar attack.

The good news was that American soldiers had repaired the generator so it was again operational.

When the base at Na Trang was reported as surrounded, no attempt was made by the enemy to capture it. It was later discovered that the manpower of the NVA teams at that time was expended, causing them to fail to capture other compounds, as well.

News from the base at Can Tho was less positive. Their generator outside of the compound walls had been sabotaged by Viet Cong sympathizers posing as Vietnamese operators. All radio traffic needed to be switched to the SSB.

When the U.S. Embassy came under attack, its personnel, as well as Marine guards and military police, managed to ward off the enemy. This was a significant achievement with long-reaching impact. If North Vietnamese leader Ho Chi Minh had captured the Embassy, American military morale and political prestige around the world would have been seriously damaged.

After the war, the efforts of those brave military personnel would inspire Bultemeyer to enter the world of law enforcement, first as a sheriff's deputy and later as a military policeman.

When sniper shots whizzed within inches of Bultemeyer's head, close enough that he was struck by shrapnel, he thought another might follow with closer designs on him, but none appeared.

At 2030 hours (8:30 p.m.) on February 1, Bultemeyer's stress level heightened when a South Vietnamese code team received a critical five-page message, detailing how a Viet Cong and NVA operation was planning to assault the Ton Son Nhut Air Base. By presenting phony documentation stating they were a South Vietnamese medical battalion with the mission of assisting American medical units on the air base with combat casualties, the enemy, dressed in medical uniforms, would disguise their real mission, which was to get on base and wreak havoc.

The saboteurs' mission would begin on February 2, 1968, at 0700 hours (7:00 a.m.).

Bultemeyer received more bad news. A breakdown had occurred between his SSB station and the receiving end with the codes. This meant the five-page message about enemy insurgents would have to be manually encoded into five-letter groups.

It would then be securely transmitted via radio to the Allies' headquarters in Saigon. Once received, it would be decoded and delivered by escorted courier to Military Assistance Command-Vietnam and the U.S. Army Republic of Vietnam at Ton Son Nhut Headquarters.

Before the message could be acted on by base air police and security forces, the message would have to be verified. From there the lengthy missive would be processed by a capable soldier with a code book in approximately 30 minutes.

Less than 12 hours remained.

Bultemeyer's tired shoulders tensed in anticipation of the work ahead of him. By concentrating on the task, he converted the message to a 22-page document in ninety minutes.

Bultemeyer then turned the document over to Major Marion Grammer at his base who ordered Bultemeyer to finally get some rest. Another operator would transmit the message to headquarters at the 525th MI Group.

It had been a grueling 72 hours. Too tired to take another step, Bultemeyer sank to the cement floor of the office into a deep sleep.

He awakened a few hours refreshed, only to be told that no more than one-fourth of the message had been transmitted to the 525th. The reason for the delay? An inexperienced operator.

Infuriated at the waste of time, Bultemeyer got on the radio to headquarters at the 525th. He didn't know the duty officer's name but asked for him to come to the radio. Bultemeyer hoped and prayed that whoever he was, he would have more experience in decoding the message than the previous soldier.

When the officer came on, Bultemeyer quickly relayed the situation, emphasizing the critical need to convert the code to normal wording and get the message to Military Assistance Command-Vietnam as soon as possible.

The duty officer, a lieutenant, protested that he didn't have time for the task.

Bultemeyer had had enough.

As the non-commissioned officer in charge of the Net, he ordered the lieutenant to take over. The lieutenant agreed, but tersely informed Bultemeyer that he would be held accountable for insubordination to an officer after things had settled down.

Bultemeyer signed off, sure that the North Vietnamese who most likely had listened in had gotten a charge out of the conversation all the way to Hanoi.

By 0640 hours, the lieutenant had completed the transmission. Only 20 minutes remained before the phony battalion was due to arrive at Ton Son Nhut.

With quick precision, the air base military police and other security personnel managed to hold off the invaders.

By the time the Tet Offensive ended on September 23, 1968, both sides claimed victory. Despite being caught off-guard, the American and South Vietnamese military almost completely eliminated enemy forces, recapturing all territory held prior to the attack and more. This neutralized the efforts of the Viet Cong and NVA.

As for the lieutenant's threat to Bultemeyer, a high-ranking officer dismissed the incident and commended Bultemeyer for his leadership.

In March 1968, Bultemeyer was re-assigned to the Ready Reserve. He returned home and in 1972 was transferred to the 601st Military Police Battalion Ready Reserve in Fort Wayne, Indiana.

In 1970, Bultemeyer married LouAnn Strader who later joined the Army. The couple became parents to four children. After retiring from the military in 2006, Fritz Bultemeyer worked for the Adams County Sheriff's Department and as a prisoner transport deputy.

"All the way through my military career God was there for me," he said. "My career always benefited from His love."

Command Sergeant Major Fritz Bultemeyer

Mike Chamness – Army

Steak, mashed potatoes, and green beans.

Mike Chamness didn't know why he and the other 98 men from A Company, 1st of 5th Mechanical for Armored Personnel Carriers were being served a better-than-normal meal. But they were not going to question it.

On that evening of February 1970, in the mess tent at Devins Fire Circle support base, 30 miles northwest of Saigon, anything looked delicious to troops accustomed to C-rations on patrols. Chamness and the others could hardly wait to dig in.

Just as they were sitting down, word arrived that a 11-man patrol was being attacked by Viet Cong a few miles away.

Dumping their food (all except the steak which the soldiers stuffed into their mouths), the men scrambled to get to the company under fire.

Finding 10 soldiers wounded, Chamness and other mechanics carried them to a chopper to be airlifted to a hospital.

One soldier Chamness transported had sustained multiple injuries, including his eyes. Chamness' insides wrenched at seeing a fellow soldier's pain.

When another soldier confessed to feeling nauseous at the sight of the grisly wounds, a third soldier carrying an M-16 in each hand, shouted, "You can't be sick! You're needed right now!"

The sight of so much carnage caused Chamness, a 22-year-old farm boy from Selma, Indiana, to wonder when it would be his turn to die.

**

After graduating from Selma High School in 1967, Chamness worked at the General Motors plant in Muncie, Indiana, until receiving his draft notice in May 1968. He wanted to be in a group that built things, such as the Navy Seabees. Growing up on a farm, he had often worked on machinery.

Note: The word 'Seabee' is a derivative of the initials *C* & *B* from the words 'Construction Battalion.'

Chamness was not drafted into the Navy but the Army. He was sent for basic training at Fort Campbell in Kentucky. When tested for strengths, Chamness' skills with mechanics caused him to be trained as a bulldozer operator in the Army Corps of Engineers.

After completing operator training, Chamness left for Vietnam on July 3, 1969. He landed two days later at Cam Ranh Bay Base on the southeastern coast.

Upon disembarking from the plane, Chamness' olfactory sense was struck by an unpleasant and unidentifiable smell that seemed to permeate the base. He came to believe it to be the combination of hot bodies, foliage, and fear.

Another discomfiting sensation was the bus that hauled him and other recruits to Cu Chi Base Camp northwest of Saigon. The bus featured wire mesh windows, ostensibly as protection from sniper shots.

Upon arriving at Cu Chi, Chamness had barely stepped foot outside of the bus when the enemy began sending mortars. He and the other young recruits stood frozen in place. Someone shouted, "Get to a bunker!" and they ran without being injured.

At Cu Chi, a base shared by the American Army and Army of the Republic of Vietnam (ARVN), the new arrivals assigned to the 25th Infantry Division received a week of basic orientation, including information on how to adapt to and survive in the new country. They also practiced getting in and out of choppers quickly.

Upon being re-assigned to A Company, 1st of 5th Mechanical as a mechanic for Armored Personnel Carriers (APCs), Chamness became one of an 11-man crew. He had never heard of the vehicle but was willing to learn.

During patrols with their APC, the crew sat on top on extended steering levers, or laterals. An APC weighed 11 tons, making it nearly impossible to blow its motor. Still, sometimes it needed repairs.

Chamness studied the machine's specifications, finding that his years of working on farm machinery helped for this vehicle needed in war.

Despite its formidability, an APC usually didn't go on patrols unaccompanied. So, when a new captain sent out a lone APC, Chamness judged him a lousy leader, especially as their radar showed 50 Viet Cong nearby. When the enemy didn't attack, Chamness suspected it was only because they believed the Americans had too much fire power.

It was sometimes debatable who was better equipped. American troops typically carried 50-caliber weapons, but Viet Cong soldiers carried AK-47s, captured from American soldiers, as well as rocket-propelled grenades.

Chamness credited a sergeant for teaching him what look for on patrol, how to shoot a M-79 grenade launcher, and other practical skills for survival in combat.

Injured soldiers received medical attention at aid stations. Courtesy Rod Maller.

When A Company's motor pool sergeant was admitted to the hospital for appendicitis, Chamness was assigned to replace him. With seven new mechanics in the unit, Chamness pulled together a team of 50-caliber gunners from among them.

Later, when a driver for a Vehicle Trac Retriever (VTR) was wounded, Chamness took his place. A VTR sported a 550-horsepower and 525-cubic engine, enabling its crew of three to use a boom extension to retrieve mines, tanks from mud, and other obstacles. One VTR was assigned to each company.

During month-long patrols, troops observed a grueling routine of guard duty with two hours on and two hours off. When given the opportunity to sleep, Chamness sank his six-foot, four-inch body into a ball inside the VTR to catch what rest he could.

Chamness worked closely with other soldiers, but he made it a practice not to become friendly with anyone. The unit had a big turnover, due to booby traps, and he didn't want to lose friends.

One night, Chamness was on duty doing maneuvers on the VTR when one of the company's 50-cal gunners radioed their captain. He had spied the enemy lying in wait ahead. The VTR captain refused to stop and the situation quickly escalated so that a chopper was called in to help.

Chamness had been in life-threatening episodes before. Now he felt terrified and prayed to God for help. The situation was resolved when the chopper fired on the enemy.

At the end of 30 days, the troops returned to Devins Fire Support Base where they ate hot cooked meals and re-supplied their food packs. One food

Chamness couldn't tolerate was powdered eggs – they appeared green when cooked.

The soldiers sometimes gave food to Vietnamese children but not canned goods. The Viet Cong were known to take cans with sharp edges and turn them into weapons.

If the possibility of becoming a casualty was not enough, Chamness received a 'Dear John' letter from his girlfriend at home. On the plus side, he and other troops were entertained by seasoned USO performers like Bob Hope.

By the first week of April 1970, activity was beginning to wind down in Vietnam. The American government authorized some troops to return home short of their one-year tour of duty.

With three months to go before completing his one-year in-country, Chamness was about to board a chopper to Australia for R&R (rest and relaxation) when his executive officer pulled him aside.

"You've been ordered to go home," he said. "You can go on R&R and then go home when you get back. Or you can leave for the States now."

Spec 5 Chamness threw aside the opportunity to visit 'The Land Down Under' and within 24 hours, he was on his way home. He was thankful to have never been injured.

After being discharged, Chamness resumed working at the GM factory in Muncie, later transferring to Fort Wayne. One of Chamness' officers in Vietnam had put his name in for a Bronze star which he received a few weeks later.

In May 1991, Mike's father expressed interest in hearing about his experiences in Vietnam. It was the first time he had asked Mike for that information. Mike promised to return in three days to talk. Unfortunately, his father died of a heart attack before Mike could return.

Mike Chamness regretted not talking to his father about his time in Vietnam. To compensate for this and other troubled thoughts he had about the war he wrote poetry about his thoughts of war.

Chamness married and became the father of two sons. When the marriage dissolved, he re-married. Today, Mike Chamness is retired and lives in Dunkirk, Indiana.

He has experienced a number of health issues, including nervous breakdowns, COPD, Parkinson's, depression, hearing loss, and PTSD. He attends counseling to help deal with anger issues.

Despite his hard times, Chamness has no regrets. "I wouldn't trade my experiences in Vietnam for anything," he said. "I left a boy and came home a man."

Spec 5 Mike Chamness (right)

Dennis Covert – Navy

During battle near Vietnam's demilitarized zone (DMZ), the crew of the USS *Saint Paul* provided gunfire support for ground troops. When ordered to general quarters, the crew immediately left their duties to report to assigned stations for battle preparedness.

Dennis Covert handled ammunition below deck, usually the five-inch guns which could fire 10 miles. The ship's largest salvo, an eight-inch gun, required 90 pounds of gun powder to fire 20 miles. The ship also handled three-inch guns.

As ground troops could request fire power at any time, the *Saint Paul*'s crew remained at general quarters 24/7. Covert wore headphones for protection from the ear-splitting noises of battle and his ears rang long after the war ended.

One thing was evident to Covert during every second of his tour of duty in Vietnam: the North Vietnamese military wanted nothing more than to demolish American troops.

**

Upon graduating from Muncie Central High School in 1965, Covert knew the American government would require him, like every male who turned 18 years old, to register for the military draft. He decided to take some control of his future and enlisted in the United States Navy.

Covert completed boot camp at Naval Station Great Lakes in Chicago and nine months as a reservist before joining active service for two years. In 1968, Covert was assigned to the *Saint Paul*. When not in battle, he worked in the ship's office and stood watch on the open bridge.

The 673-foot heavy cruiser had been commissioned in February 1945 and sent to the Pacific where it fired on Japanese targets during World War II. When Japan surrendered in August 1945, the ship, as part of Admiral Bill 'Bull' Halsey's Third Fleet, sent the last naval salvo on that country's homeland. On September 2, 1945, the *Saint Paul* sailed to Tokyo Bay for the formal surrender ceremony.

The ship continued its involvement with battles in the Korean Conflict, firing the final naval salvo on July 27, 1953. By 1964, the *Saint Paul* was one of the few World War II cruisers still in commission. The following year, she was used in the filming of the motion picture 'In Harm's Way' starring John Wayne.

On September 1, 1967, before Covert arrived in Vietnam, the *Saint Paul* was hit by an enemy shell on her starboard bow. None of the crew was injured and her engineers repaired the slight damage, enabling the ship to continue her mission.

The ship sailed to Japan, Taiwan, Thailand, and the Philippines. The sight of American military troops generally was welcome in most ports because it meant money for restaurant and shop owners. But sailors were not always safe. In Thailand they were ordered not to go off base alone, due to the presence of communist insurgents.

As the *Saint Paul* sailed through the South Pacific, she encountered typhoons lasting for days. The crew rode them out with all hatches sealed down and no one allowed on deck, due to the danger of being washed overboard. Covert's unease during the storms was intensified with episodes of seasickness.

In October 1969, Petty Officer Third Class/ E4 Dennis Covert was discharged. He used the GI bill to enroll at Ball State University in Muncie, Indiana, where he earned a degree in business administration. At 25 years of age, Covert was only a few years older than most of his classmates. Due to the war, he felt older by decades.

After graduating, Covert was hired as a stockbroker for Merrill Lynch in Fort Wayne, Indiana, a position where he remained for 43 years.

In 2010, Covert became involved with Honor Flight of Northeast Indiana. This was the local hub of a non-profit national organization that flies American military veterans on all-expense-paid trips to Washington D.C. to see war memorials. Covert sponsored a trip and served as president for five years. He continues to volunteer as a guardian for veterans and in other areas.

He is a former historian for the USS *Saint Paul* Association and a current member. "The Fighting Saint", as the ship was nicknamed, earned 18 battle stars for combat operations during her three conflicts. She was one of the last all-gun cruisers and fired more rounds than any ship in the history of the United States Navy before she was decommissioned in April 1971.

"I am just one of a whole generation of people who has served our country in the military," said Covert. "We did our part. We were not heroes, but it made us tough. I encourage young people today to consider military service. It's a tremendous maturing experience they will have forever."

For more information about Honor Flight of Northeast Indiana contact http://hfnei.org./.

Petty Officer Third Class/ E4 Dennis Covert

Richard Dawson -- Marine

At daybreak on June 18, 1968, 10 Huey helicopters hauling Corporal E4 Rick Dawson's Marine unit of 500 soldiers landed at Marine Corps Combat Base at Con Thien. The American Army Special Forces camp was two miles from the DMZ. After transitioning from the Army to the Marines, it had seen fierce fighting.

The Marines' objective was to march from Con Thien to the city of Da Nang where enemy forces lay in wait. The troops were assured of air and artillery support, as well as B-52 bombing, during the operation codenamed Pegasus.

When the 10 Hueys, each carrying 50 soldiers, landed at Con Thien, all was quiet. Dawson's chopper had been first to land. His group dug foxholes as the other choppers landed.

Suddenly, shelling fired around them. It didn't take Dawson or the others long to realize they were in the midst of an ambush.

Dawson scrambled to safety. A pain behind his left ear felt as though he had been hit by a baseball bat. The injury deafened him and he could not see. He realized he had probably been struck by a shell.

Uncertain which direction led to safety, he inched forward while shouting for help.

When his hand hit air, Dawson tumbled, falling into a hole with a hard surface. He guessed it was a bomb crater. Stunned, Dawson lay still, wondering which would come first -- capture or death?

**

Richard Dawson had been born and raised in Oakes, North Dakota. After graduating from high school in 1966, he enlisted in the United States Marines.

Like most teens, Dawson had not known much about the war being fought in Vietnam, the small country halfway around the world, when he heard it discussed on the television news. But he was determined to serve his country, no matter the cost.

After completing basic training at Marine Corps Recruit Depot San Diego, he left on an American Airlines 747 jet full of other recruits in January 1968 for Vietnam.

The first thing that struck Dawson upon exiting at the city of Da Nang was the intense heat. He guessed the temperature to be well above 100 degrees and imagined it would feel the same as walking into a furnace.

Dawson didn't have long to dwell on his discomfort when a whistle sounded followed by an explosion. "Spread out!" someone yelled.

He and the other young men recently arrived in the country dove behind buildings for protection from enemy fire.

Dawson escaped injury that day, but was sad to hear the officer who had issued the warning had died while trying to save the newest recruits. Dawson would never forget the sacrifice of his unknown savior.

Dawson's first duty station was Khe Sanh, a village 14 miles south of the DMZ, six miles from the Laotian border. Surrounded by 15-foot elephant grass, the soldiers held rifles over their heads while conducting patrols in the area.

Dawson felt fortunate that he had not been there six months earlier when soldiers at the Khe Sanh garrison were attacked by forces from the People's Army of North Vietnam (PAVN). For 77 days, U.S. Marines and their South Vietnamese allies fought off a massive artillery bombardment, making the siege one of the longest and bloodiest battles of the war.

As Dawson and many other soldiers would find, the enemy was not always in human form.

The American government sprayed the foliage in the area of Khe Sanh with herbicide, setting in motion actions which would lead to devastating results for the health of soldiers serving there.

The soldiers could not afford time or energy thinking about their futures. They had to focus on surviving each day — an arduous task when the entire country seemed to be under siege and no one could be trusted.

Dawson lay in the bomb crater what seemed to hours, unable to move and hurting all over. Finally, he crawled to where a corpsman could administer aid to him. Two hours later, Dawson was airlifted to Da Nang Hospital.

Doctors found a bullet hole above Dawson's left ear exiting above his brain. Two inches difference would have paralyzed or killed him. Dawson was flown to Guam Naval Hospital where he underwent surgery on his skull.

After a month of recovery and therapy, the medical community, not seeing improvements, sent Dawson to Oakland Naval Hospital in California for further treatment.

Before leaving Asia, Dawson learned the sad news that 42 soldiers from his chopper had been killed at Con Thien. No air support ever arrived.

Despite the discouraging progress of American troops in Vietnam and his extensive injuries, Dawson felt lucky to be going home in one piece.

But, it was 1970 and in America depressing actions due to the war were taking place. War protesters called Vietnam War veterans "baby killers" and other derogatory names.

Hearing about the Watergate scandal, Iran Hostage Crisis, amnesty for "draft dodgers" (nickname for men who went to Canada rather than serve in the military) caused Dawson more pain than being injured by the North Vietnamese.

Eight years passed before anyone thanked Dawson for his military service.

Over the years, Dawson tried to work, but headaches, nerve damage, heart problems, and post-traumatic stress disorder made it difficult to hold a job. His health problems continued through the decades, causing some professionals to believe his afflictions were due to the effects of Agent Orange.

Dawson married and he and his wife, Linda, became parents to four children. Despite his afflictions from his military service, Dawson tried to instill a sense of patriotism in his brood, the result being two sons serving in the Marines.

Rick Dawson is a long-time member of Pride & Purpose Detachment Marine Corps League in Fort Wayne, Indiana and has marched with a local American Legion's Honor Guard.

"I'm honored and proud to have served my country in Vietnam," he said. "But I do wonder about the future of our troops."

Corporal Richard Dawson

Mike Dean – Army

On August 5, 1968, Spec 4 Mike Dean was not supposed to be in the air.

As crew chief for Huey helicopters, Dean performed daily maintenance on the state-of-the-art aircraft that could carry two pilots, two gunners, and up to six passengers.

Note: Huey is a nickname for the acronym assigned to the "helicopter utility" (HU) aircraft.

In addition to working on the Hueys, Dean loved flying in them. When gunners were short, he often volunteered to man the M-60 machine guns during missions.

Though flights over enemy territory were dangerous, Dean relished the action. By frequently replacing crew members unable to go on missions, he had accumulated 150 flight hours.

Even though August 5, 1968, was Dean's scheduled day off, he had again volunteered for the day's mission which was to haul grunts (slang for American infantrymen) to a combat area.

As the chopper approached the landing zone, Dean, strapped into his seat next to the open door, provided cover as troops dropped to the ground with their guns.

At one point he felt stings to his left arm and hip. Suspecting he had been shot, Dean said nothing until the pilot safely landed the aircraft away from the battle. When the pilot learned of Dean's suspected injuries, he became alarmed and quickly flew him to a hospital at Cu Chi where Dean underwent surgery to remove shrapnel from an AK-47.

**

Born in Detroit, Michigan, Mike Dean had grown up in Milford, graduating from Milford High School in 1966. Believing he would be drafted and wanting to have some control of his military assignment (he hoped to be a helicopter mechanic), he had enlisted in the Army.

In January 1967, Dean was thrilled to see his wish come true as he was sent to helicopter mechanic school at Fort Rucker in southeast Alabama. The base was the center of army aviation and Dean received additional instruction in crash rescue procedures.

Upon arriving at the base at Cu Chi on New Year's Day 1968, Dean, having completed his training, expected to be assigned as a crew chief. He was less than thrilled to be re-assigned as a clerk for an office position when his background revealed completion of a typing course in high school.

After four months, Dean asked for a re-assignment. At Tay Ninh, a city 60 miles northwest of Saigon, he was re-assigned as crew chief with the 187th Assault Helicopter Company (AHC).

Army doctors removed the bullet from his elbow and Dean received 11 metal stitches and ten more in his hip for the injury there. For recovery He was taken to an American army hospital at Camp Zama in Japan, 20 miles southwest of Tokyo.

Knowing how upset his mother would be at the news of his injury, Dean convinced the military to let him write to her, detailing what had happened. Once his letter writing was up-to-date, Dean was left with little to do. Remaining idle was a challenge for the soldier who preferred activity.

A bright spot occurred at Camp Zama with a visit from Debra Barnes, winner of the Miss America Pageant 1968. She and her USO troupe visited soldiers throughout the country.

In October 1968, Dean returned to the United States for active duty at Hunter Army Airfield in Savannah, Georgia. He remained there until discharged in January 1970 at the rank of Spec 5 (E5).

Like many Vietnam War veterans, Dean discovered returning to civilian life almost as challenging as serving in a war zone. While working at a department store, he was approached by two male co-workers. When they asked him why he had gone to Vietnam, Dean replied, "For you."

Their reply: "We wouldn't have done it for you."

In May 1970, the same co-workers, who were members of the Ohio National Guard, were activated to serve at a protest at Kent State University. When members of the Ohio National Guard opened fire on a crowd gathered to protest the Vietnam War, they killed four students and injured nine. The nation was outraged.

When the co-workers returned to their jobs, they avoided talking to Dean. He suspected they had a new perspective of military service.

Dean worked many years for the U.S. General Services Administration. Retired, he lives in Clarkston, Michigan where he is a member of a Community Emergency Response Team and stays involved with his local veteran center, speaks to school classes, attends church, and visits with family (he has one daughter) and friends. "I was proud to serve my country," he said. "I was glad to go and do something to help."

Spec 5/ E5 Mike Dean

Randy Harnish — Army

In October 1969, Spec 5th Class Randy Harnish loaded his jungle boots, fatigues, and weary body aboard a Huey parked on Hawk Hill, about 30 miles south of Da Nang. He had just completed a six-month assignment with the U.S. Army's Americal Division. The Americal Division (23rd Infantry) was formed from elements of American troops.

Known for its jungle fighting lineage from World War II, in Vietnam it became the largest infantry division, covering a variety of terrains and more territory than any other division.

The pilot carried Harnish west to his next assignment with the 3rd Battalion, 16th Artillery Complex. It was a special forces camp in the Central Highlands near the village of Tien Phuoc. In the mountainous region Harnish would continue with his duties involving meteorological ballistics (study of increasing the range and lethality of munitions by applying environmental effects).

When the chopper pilot put Harnish down on a hill, a small Vietnamese boy stood to the side. He indicated Harnish should follow him and Harnish did so, though warily. Sometimes the Viet Cong used children to deceive the Americans.

During the walk, the child fell several times. When Harnish offered to carry him, he declined and resumed walking.

Upon arriving at the fire base, Harnish settled in among the 85 soldiers who were members of a heavy artillery battery. They nicknamed themselves the 'Rolling Thunder Battalion.' Curious about the child who seemed to be living at the base, Harnish gleaned what little was known about him.

**

Little Lou, as the soldiers at the base named him, had wandered into the special forces camp three months earlier, weak from malnutrition and barely able to stand. As the medical staff treated him, they assessed his age to be around seven years, though they believed the child had been born with a disorder that stunted his growth, possibly dwarfism.

Lou was part of an indigenous group called the Montagnards (French for 'people of the mountain'). The Montagnards were known to reject babies with birth defects. As Lou didn't speak English, the soldiers surmised the Viet Cong had raided his village, killing everyone they believed had been sympathetic to the American cause. Somehow the child had escaped and survived alone in the jungle for a week.

Harnish joined the other men at the fire base in caring for Lou, providing him with food, a clean place to sleep, medical care, and, most importantly, love. They taught him to speak English – or what they thought he needed to survive. While an American teacher might have lamented the slang, colloquialisms, even profanity that made up Lou's talk, he learned to communicate well with his caregivers.

It wasn't long before Lou was smiling and laughing. He took an active part in daily life at the fire base, issuing the morning wake-up call, and greeting each new arrival.

One of Lou's favorite events was the delivery of care packages to soldiers. He followed the recipients to their bunkers where he helped retrieve the goodies nestled within. Even crumbs delighted the child.

Harboring Lou was breaking every rule in the Army manual. That was what made it so enjoyable to the soldiers. "We thought that maybe during the war no one wanted and few understood that we could make a difference in someone's life," said Harnish.

Their enthusiasm dimmed when they thought of what might happen to Lou when the war ended. The soldiers would go home, which was something they all dreamed of daily, leaving Lou again on his own.

As a child struggling to survive alone in a war-torn country, his chances of survival seemed abysmal.

In the evenings when the soldiers sat around talking about their lives back home, Lou listened to stories about big homes, fast cars, paved streets, grocery stores, movie theaters, and best of all, candy and toy stores. He desperately wanted to be a part of that world.

Each time a soldier received orders to leave, Lou clung to his uniform, begging to accompany him to the States. The soldiers, most of whom were 18 to 21 years old, sadly shook their heads against his pleas. It was against Army protocol to do such a thing.

Many, including Harnish, wished they could do something. "Lou had gotten under my skin and I could only hope he would survive the war and afterward," he said.

Harnish was scheduled to leave Vietnam in early May 1970. But, on March 24, he was one of approximately 5,000 soldiers who, due to President Richard Nixon's efforts to reduce America's involvement in the Vietnam War, received an early departure.

Upon receiving notice, Harnish quickly packed his duffle bag, ran to the top of a hill where a supply chopper sat waiting to begin the trip home, and climbed aboard.

Thirty-six hours later, Harnish was sitting in a new GTO at Bummie's Root Beer Stand in his hometown of Bluffton, Indiana, eating hot dogs with a girlfriend.

While Harnish's quick departure was a mixed blessing (his trip back to Indiana was happily uneventful), he would often wish in the coming years that he had taken time to say goodbye to friends, especially Lou.

Had the child with so many negatives in his life–ill health, history of neglect, no family or home-- remained safe after the war?

In October 2010, Harnish received a phone call from Dick Harnley of the 3rd Battalion, 16th Field Artillery Regiment. The unit's former executive officer was organizing a reunion and contacting members of the Rolling Thunder Battalion.

Harnish, who had become a business owner, looked forward to getting together with his unit. He asked if Little Lou's name was on the roster and was overjoyed when Harnley replied that it was.

Lou had lived in the States for many years and planned to attend the reunion. Harnley provided Lou's phone number to Harnish who immediately called him. Lou answered and seemed equally happy to hear from one of his old friends from the Rolling Thunder Battalion.

He lived near Seattle. At Harnish's further questioning about how he had gotten to the States, Lou explained that he had remained with the American military until April 4, 1975, when he was taken to the Saigon airfield. The American military was leaving the country. It was thought to be a race against the clock before Saigon fell to the North Vietnamese.

The American government had arranged to fly thousands of Vietnamese children thought to be orphans to the United States. Government officials would find them new homes. Lou was among the group of children. It seemed his dream of living in America would come true at last.

Processing hundreds of orphans was laborious, due to a lack of interpreters. With his knowledge of English, Little Lou volunteered to assist.

When it was his turn to board the C5A plane that would fly him and other children first to Clark Air Force Base in the Philippines and then to the States, Lou encountered a glitch. He lacked two required immunization shots.

An adult hurriedly drove Lou to a hospital where his shot record was updated. By the time they returned to the airport, the plane Lou was to be on was taxiing across the tarmac.

The little boy could hardly be consoled as he watched the giant aircraft rise in the sky. It seemed his dreams of getting to the United States were dashed.

Then, to the horror of the crowd watching in the airport, the plane teetered in the air before crashing on the runway. Of the 328 people on board, 138 died, 70 of them children.

The cause for the explosion was later determined to be a structural malfunction.

Soldiers use yellowish smoke to signal locations. Courtesy Rod Maller.

The following day Lou was placed on a flight that successfully carried him and hundreds of others out of the war-torn country. Three weeks later on April 30, 1975, the North Vietnamese took control of the city.

Before leaving Saigon, Lou was interviewed by American newscaster Bill Plante. When Lou's story aired in the States, many people responded with the desire to adopt him. Among the group was a dairy family named Arestad from Ferndale, Washington who had already adopted 14 children from outside of the United States. American officials chose the Arestads to be Lou's new family.

Lou's name was officially changed to Robert Lou Arestad, but to friends and family he was still Little Lou.

The child who had been isolated, starved, and neglected during much of his childhood thrived in his new home and country. He graduated first from Ferndale High School in 1984 and then college. He worked 10 years in the banking field and started a number of businesses, including an accounting firm, commercial construction, and home mortgage venture. He sponsored a charity golf tournament to help homeless children.

"I felt God has blessed me along the way and kept me alive for a purpose," he told Harnish. In a later email Lou added: "My charity work is in part a tribute to the men of the 3rd of the 16th who I considered my extended family."

"I was an orphan and you took me in. I was hungry and you fed me. I was cold and you provided a warm blanket. I now help other children along the way to build their self-esteem and create motivation, for they are our most precious resource. They are the future captains of industry and leaders of the free world."

At the military reunion Harnish and the other veterans of Battery B greeted each other, happily recalling the bygone era. The joyous meeting seemed odd to Harnish, since 45 years ago when they had first been together, nothing had seemed funny.

A highlight occurred when Major General Tom Lightner, former Deputy Commander of Western Forces, now retired, presented Lou with a special designation as honorary member of Battery B, 3rd Battalion, 16th U.S. Artillery Regiment.

As Lou had been with the Rolling Thunder Battalion for several years, while the rest of the men had served one-year rotations, they believed he deserved the designation.

In March 2016, General Lightner called Harnish again, this time with bad news. Lou Arestad had died on January 24th from heart issues. His obituary listed his age as 53. Lightner asked Harnish to deliver the eulogy in Bellingham, Washington.

On February 8, 2016, more than 400 family and friends gathered for Lou's celebration of life service. Six veterans from his old unit attended.

General Lightner spoke first, then Bill Plante spoke via video hook-up. Plante said he had been a young, 'wet-behind-the-ears reporter' when he interviewed Lou in the final days of the war. They had become friends and stayed in touch after Lou's arrival in America.

Plante added that among the people he had interviewed during his 50-plus years at CBS, no one had been more interesting and inspiring than Lou Arestad.

Harnish gave a eulogy, then nieces and nephews stated how much they had loved Uncle Lou as the family 'go-to' guy.

Today, when Harnish reflects about his time as a soldier, Lou is included in his thoughts. "The war in Vietnam has long been regarded as the most unpopular war in America's history. However, I believe a handful of soldiers will go to their graves believing they made a difference in a young boy's life ... and I'm proud to be in that number.

Randy Harnish and "Little Lou" Arestad at a military gathering in 2010.

Arley Higginbotham -- Army

Second Lieutenant Arley Higginbotham and members of 2nd Battalion, 47th Infantry, 9th Infantry Division, Company C stared into darkness, waiting for the enemy to attack. It was January 1968, and the unit had been assigned to an ambush site, along with members of ARVN.

Suddenly, a loud noise erupted. The trigger-happy American soldiers fired, ceasing only when Higginbotham frantically gave the signal. He had discovered the cause of the disruption -- an ARVN's portable radio turned on at full blast.

The company's position was now compromised. As they began receiving fire from the enemy, a 50-cal machine gunner on Higginbotham's track was hit in the right shoulder. The injury spun him to the right, too fast for him to release the trigger of his gun and a stray round was loosed at a neighboring track. It hit the machine gunner, a sergeant, the round leaving a large hole in his back.

Higginbotham's crew scrambled to provide cover, while calling for a med evac chopper. Waiting for it to land, Higginbotham held the young man's head in his hands, stricken at seeing his backbone blown out through his stomach.

"I'm going to make it, Tuey, right?" he whispered.

Higginbotham told him yes, as the injured man took his last breath.

The flashing of lights in the night, thumping of the Huey's blades, the young soldier dying in his arms -- all combined to create moments Higginbotham would never forget.

**

Arley Higginbotham enlisted in the Army after graduating from Barberton High School in Barberton, Ohio, in 1965. He attended Army maintenance school in Fort Knox, Kentucky, before leaving for Vietnam.

Upon arriving at Tan Son Nhut Airbase in early January 1968, Higginbotham and other new recruits disembarked from the plane as enemy mortars exploded on the runways.

Higginbotham had been trained as a battalion motor officer and he expected to be placed in that position in his new station in Bien Phuoc. Instead, he was assigned as a frontline platoon leader in 2nd Battalion, 47th Infantry, 9th Infantry Division, Company C. They were nicknamed "Bandito Charlie."

Higginbotham was flummoxed. He had never worked with a platoon in combat situations and other than performing maintenance on the equipment, he had no experience with M-113 APCs.

He called a meeting and informed the soldiers of the mechanized infantry company, all of whom were between the ages of 18 and 21, that he had no experience. He depended on their knowledge of the area. Higginbotham added that he would do his best to get them home alive.

On January 31, chaos ensued in Saigon and other major South Vietnamese locations as 85,000 troops of the North Vietnamese launched a series of surprise attacks, including at Ton Son Nhut Airbase. Higginbotham wondered if he could keep his promise.

An intel report stated that the Viet Cong was going to attack during the celebration of the New Year Festival called Tet.

On the holiday which in past years had been observed as a day of truce between South Vietnam and North Vietnam and their allies, the Viet Cong had hit hard with small arms and mortars.

A door gunner prepares to fire from a helicopter. Courtesy Rod Maller.

Higginbotham's unit held their positions and survived with no injuries.

The next day, Charlie Company was sent into Saigon, another area under duress. Moving down one of the main streets, they received a call that Alpha Company was pinned down in an alley in the opposite direction taking heavy fire. Many were reported wounded and one of the tracks was on fire.

When the company reversed to help, Higginbotham's track became the lead. As the driver drove through the streets, Higginbotham nestled on top of the 11-ton APC, heart pounding, finger poised on his gun.

They found the alley with the burning track. When an American officer refused their entrance because Charlie (the enemy) had RPGs that he thought could destroy the American tracks, Higginbotham relayed the information to his officer who told him to go in anyway.

As they surged forward, they pushed the burning track at an angle, causing it to spin awkwardly and prevent the rest of the company from following and lending support.

Higginbotham's unit saw a huge hole blown in the wall of the alley with a cemetery on the other side. Presumably the enemy was dug in nearby.

Now Higginbotham's unit was not only exposed to the enemy but alone.

Higginbotham decided to take the fight to the enemy. He set up everyone with a machine gun, equipping a M-79 man, as well.

Their charge into the graveyard with so much fire power surprised the Viet Cong who hid inside of tunnels. The APCs circled back to the hole in the wall and drove to the location of the wounded.

Higginbotham ran with desperation, carrying wounded soldiers toward the APC. When the Viet Cong concluded no help was coming to aid the Americans, they resumed shooting. Higginbotham watched in dismay as bullets narrowly missed his driver's head as they made multiple trips.

Retrieving the last wounded and while still under fire, the unit fled the alley, calling in gunships to clean up.

Later, Higginbotham received a Bronze Star and his driver the Silver Star for their actions that day.

Note: The Bronze Star is awarded to members of the United States Armed Forces for heroic achievement, heroic service, meritorious achievement, or meritorious service in a combat zone.

The Silver Star is the third-highest military combat decoration that can be awarded to a member of the United States Armed Forces. It is primarily awarded for gallantry in action while engaged in action against an enemy of the United States and while engaged in military operations involving conflict with an opposing foreign force.

A few weeks later, Higginbotham was in a battle in a rice paddy outside of Saigon when his company was flanked with heavy rounds of gunfire.

Higginbotham called for a helicopter strike, while he and others in the company continued firing at the enemy. He ignored a pain in the top of his thigh, though when the helicopter arrived and wounded were placed inside, soldiers around him urged Higginbotham to seek medical care at the evac station.

Higginbotham objected, believing his injury to be insignificant, based on the pinkie finger-sized hole in his pants. But when he began to feel weak, he finally agreed to go.

During an examination inside the inflatable hospital tent, doctors found Higginbotham's injury closer to the size of a silver dollar.

When enemy forces mortared the hospital, causing the tent to deflate, Higginbotham felt more terror on the surgery table than in the midst of battle. He yearned for his weapon.

Higginbotham survived the injury and spent five months recuperating at a hospital in Saigon. The long days were broken up by a visit by actor Sebastian Cabot who played Mr. French on the popular television show *Family Affair*. Against the advice of many who said it was too dangerous, Cabot traveled to various hospitals in Saigon, visiting wounded soldiers during the war. "It was a thrill meeting him," said Higginbotham.

By the time Captain Higginbotham was discharged in 1970, he had received two Purple Hearts -- during another battle, he had sustained yet another gunshot, though less serious than his first injury.

Back in the States, Higginbotham married and had a family. He served several decades as a preacher in the Midwest.

"The memories of war are no longer painful," he said. "God has healed that. That's the great thing about God. He can heal memories."

Captain Arley Higginbotham

Jose Huerta — Navy

In December 1969, Jose Huerta's first order out of corpsman school startled him. He was to check the presentation of a deceased naval officer in his dress uniform. The officer's remains were to be sent home for burial and the Navy wanted him ready for the family.

This stark reality of the consequences of life in the military quickly introduced Huerta to his chosen profession. As a corpsman (Navy's term for medic), he knew each day would be a struggle with life and death.

**

That same struggle had existed within the Huerta family since the first decade of the twentieth century when his maternal and paternal grandparents had fled Mexico as a result of violence from the revolution. They settled in Peoria, Illinois, and Kansas City, Kansas. Jose's father, who served in the Army in World War II, met his wife – Jose's mother -- when stationed at an Army base in Peoria.

After marrying, they became parents to five children, including Jose born in 1947.

In 1964, Jose, 17, like most teens, wondered about his path in life. He recalled years earlier when he had stood in line with his aunt to receive cheese and bread distributed for free to people of limited income by the American government. Jose had been embarrassed to accept the items and told himself he would someday be on the giving side.

As a teen, he also recalled feelings of pride as his uncles and his father had worn their military uniforms and recited stories of adventure and travel in service to their country.

With those memories in mind, Jose took to heart the words of the Navy recruiter who came to his school to tell students about how the Navy could provide a career, good income, and travel experiences.

Jose Huerta decided to join the Navy. As he was 17 and underage, he had to have his parents' written permission to enlist, which they provided.

Jose was especially interested in training as a corpsman. During much of his childhood, his mother had suffered from tuberculosis, even being admitted to a sanatorium for intense treatment. Confronted with this and other family illnesses, he wanted to learn how to manage and assist in providing care to people in need of medical attention.

In 1965, Jose graduated from Spalding Institute, a private Catholic boys school in Peoria. Though he received a deferral from active military service to attend Bradley University in Peoria, he completed annual two-week training periods with the Navy, including basic training, assignment to a personnel carrier in Benton Harbor, and rehabilitative care training at Great Lakes Naval Hospital.

After graduating from Bradley in May 1969 with a degree in Spanish Secondary Education, Jose received orders to report to Balboa Naval Hospital, San Diego, for a four-month formal corpsman school.

Balboa was the largest military hospital in the nation with 4,000 beds. Its staff treated mostly soldiers seriously injured in Vietnam and many patients received prosthetics and rehabilitation. Huerta was appointed adjutant of a company of 60 fellow trainees.

Ramon Huerta, Jose's younger brother, was completing his third in-country deployment with the Marines. He had been involved with serious fighting, including the Tet Offensive. Jose Huerta worried about Ramon and hoped someone would care for him, should he be injured, as Jose was caring for other soldiers.

In December 1969, Jose Huerta was transferred to Boston Naval Hospital where he was assigned to an intensive care unit. Corpsmen performed similar duties of civilian licensed practical nurses, while registered nurses – usually female -- supervised.

In October 1970, Jose Huerta sailed from the West Coast on the USS *John Hancock* (CVA 19) to the South Pacific. The *John Hancock*, an attack carrier built in World War II and named for the president of the Second Continental Congress, boasted 13 decks and a crew numbering 3,450.

The medical department consisted of a doctor, surgeon, and 20 corpsmen. Huerta sutured wounds, managed clinics aboard ship, helped in surgery, provided emergency care, and stood medical guard on an upper deck when aircraft were in the air and when the ship was being re-supplied.

When asked to care for injured and even deceased soldiers, such as the naval officer, Huerta considered it a privilege not to be taken lightly as he began his service in Vietnam.

He learned the importance of appearing calm when dealing with emergencies. Inside, however, his thoughts were often chaotic.

He was especially terrified at night when sharing a room three decks below the water line with 19 sailors. With bunks stacked four high, the proximity, plus the ship's noises and clatter of feet on the metal stairs as sailors returned from shifts added to his sleepless nights. 'What if the ship gets in trouble?' he asked himself. 'Will I get out?' His worries and stress of his job caused Huerta to feel like a volcano was happening in his head.

Then, one day, a thought struck home. If the ship went down, he would probably go down with it. At that point Huerta thought no more about survival and proceeded with his duties.

As the *John Hancock* approached Vietnam's east coast, it stationed itself three miles from the South Vietnam's coast in open waters. Americans called the strip north of the 19th parallel "Yankee Station."

Note: The 19th parallel was the Vietnamese Demilitarized Zone established as a dividing line between North and South Vietnam. The zone ceased to exist at the war's end in 1976.

As this was neutral territory, American military officials believed the location would prevent the ship from being targeted by the enemy.

Each day American aircraft lifted from the *John Hancock*'s deck for strikes against North Vietnamese missile and antiaircraft sites. From the sick bay, Huerta and other medics watched via closed-circuit TV. Navy pilots were some of the best in the world. Still, it was a challenge to negotiate a carrier's runway and pitch.

If an emergency occurred, Huerta grabbed his bag of medical supplies and climbed the ladders to check for injuries.

Sometimes a crash caused the deck to be set afire. During the two minutes it took Huerta to arrive at the scene, the ship's fire tender crew usually had the flames under control, leaving medical personnel to tend the wounded.

Once, a pilot miscalculated his landing and his body was thrown from the plane. "We found him six decks up," said Huerta. "It was too late to help him."

In January 1971, Huerta was informed that the ship's X-ray technician was leaving in three weeks. Huerta was ordered to take over his duties. He learned how to take X-rays, and develop them in dunk tanks before reading them. He also learned to apply casts for broken arms and legs.

The *John Hancock* remained at Yankee Station until May 1971 when she returned to the United States for repairs.

After three months in dry dock, the ship returned to Vietnam. Huerta, who had re-enlisted for a second tour, remained a member of the crew.

By the time his second deployment ended in March 1974, Jose Huerta had reached the rank of E5 Hospital Corpsman Second Class. He was discharged, having served 10 years in the Navy Reserves which included three years of active duty.

Huerta returned to college to earn a bachelor's degree in nursing at Ball State University in Indiana. In Peoria he filled nursing positions at Proctor Hospital, St. Francis Hospital, and Pontiac State Prison.

In the late 1970s, Huerta moved to Parkview Huntington Hospital (then Huntington County Hospital) in Huntington, Indiana, to work and would remain there for 23 years.

Huerta married in 1982. He and his wife, Anita, fostered 53 children, adopting seven before her death in 2012. Huerta continues fostering today.

During Ramon Huerta's service in Vietnam, he was exposed to Agent Orange. He died in 2021, due to the COVID virus complicated by the chemical exposure in Vietnam.

Upon arriving home from Vietnam, Jose Huerta didn't feel his service was appreciated by people outside of his family. He held back from talking about his experiences.

In recent years, he has shared his story of military service as Americans began recognizing the service and sacrifice of Vietnam veterans.

"It meant a lot to us to serve our country and contribute to its peace," he said. "We worked as shipmates towards the same goal, no matter our differences. We were thrown together but finally developed a mutual trust to get the job done."

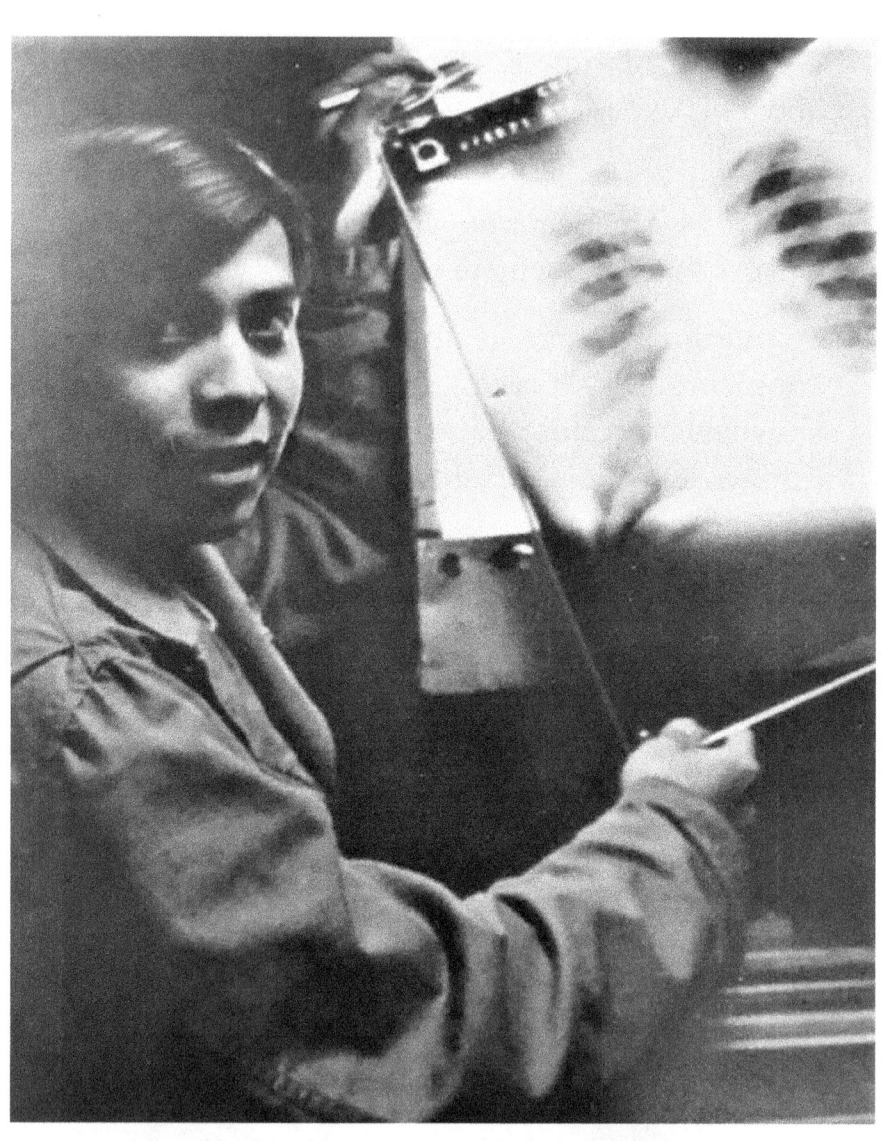
E5 Hospital Corpsman Second Class Jose Huerta

Lanny Idle – Air Force

Boom! Boom! Boom!

Mortar fire shook Sergeant Lanny Idle awake in his barracks at Phan Rang Air Base. Though the base in southeast Vietnam was not often under attack, the men stationed there knew when it occurred, they should don flak vests and helmets and quickly head for a nearby bunker.

Idle and the other men sprang from their beds and headed for the bunker. As they hunkered inside, waiting for the shelling to stop (30 minutes) and the all-clear be given, the men looked at each other – and laughed.

No one had remembered his helmet or flak vest.

The incident was humorous, and yet it emphasized to the young men of the need to be diligent and think about safety at all times.

**

Idle was born in Warsaw, Indiana, in 1944. Other than uncles who had served in the Korean War, he had no connection with the military.

That changed when a brother-in-law took him as a teen to Grissom Air Base in Peru, Indiana, for an air show. The exposure to planes prompted a desire in Idle to learn more about aircraft.

After graduating from South Whitley High School in 1964, Idle enlisted in the Air Force. The Vietnam War had kicked off and he wanted to get in the branch of his choice.

Upon completing basic training at Lackland Air Force Base in San Antonio, Idle, like all recruits, was tested for his strengths and aptitude. He was found to be strong in mechanics and trained for seven months at a school at Chanute Air Force Base, 130 miles south of Chicago.

Throughout its opening in 1917 to closing in 1993, Chanute's primary mission was as a technical training facility. Idle trained as a line repairman working on aerospace ground equipment (AGE).

An AGE crew maintained equipment that supported aircraft on the ground. They also ensured crew chiefs had air compressors, heaters, air conditioners, generators, and gas turbine engines to be used for maintenance on planes.

In June 1965, Idle was assigned to the 4453rd Training Squadron for pilots. At Davis Monthan Air Force Base in Tucson, Arizona, his duties continued with aerospace ground equipment, specifically on F-4s.

Fourteen months later, Idle flew to his first overseas assignment at Ubon Royal Thai Air Force Base, 40 miles west of the Laotian border. He and a crew of 20 were assigned to the 8th Tactical Fighter Wing working on AGE equipment.

From 1961 to 1975, the American and Thai governments worked together on Thai bases to deploy allied aircraft for the war effort. American planes taking off from Ubon, Takhli, Korat, U-Tapao, Don Muang, and Udorn were responsible for the majority of American Air Force air strikes over North Vietnam.

At Ubon, Idle worked eight-hour shifts, usually during the day, occasionally at night. The department operated 24 hours a day.

His AGE shop operated inside a tent. During the day, hot compressed air held up the rubberized fabric. At night, cooler air from outside caused the tent to collapse, dousing workers within with condensation. He was grateful when, two months later, a metal building was constructed.

Americans were not the only military at Ubon. In May 1962, the Royal Australian Air Force (RAAF) sent a detachment of eight aircraft. Their mission was to assist Thai and Laotian governments in actions against communist insurgents during the early years of the war.

The Americans and Australians performed joint exercises and provided air defense for the Air Force attack aircraft and bombers based at Ubon, although Idle's AGE shop didn't work on their planes.

In August 1967, when it came time for Idle to decide about re-enlisting and choosing his next assignment, he didn't hesitate. He wanted to go to Vietnam.

He knew some people would say he was young and dumb to want to go to the war-torn country. But, Idle had heard of protesters against the war in the United States. He didn't want to go home.

Idle's orders cut for Phan Rang Air Force Base included a layover at Cam Ranh Bay Air Force base. The following day he would catch a hop to Phan Rang with the 35th Combat Support Group.

That night no shelling occurred on the base. But another event within the barracks shook Idle.

From his bunk he could hear someone crying.

Idle was a single man with friends and family who looked forward to his return. He realized as never before how difficult it was for married soldiers to be separated from a spouse and possibly children. Idle tried to shut out the sounds and sleep.

At Phan Rang, Idle trained for mechanical and other processes on F-100s. Pilots flew sorties (bombing missions) with F-100s over target areas of the Mekong Delta and Ho Chi Minh Trail running through Vietnam. American gunships destroyed trucks, attacked enemy encampments, ammunition dumps, and other ground targets.

Monsoon season began in October, rain pouring nearly non-stop through early December. When water covered the flightline, ditches had to be hastily built for it to escape. Idle felt lucky to be in a barracks as three weeks before his arrival, soldiers had lived in tents.

While at Phan Rang, Idle focused on getting rank. He passed correspondence courses at the 5 level maintenance. Later, in the Air National Guard, he attained 7 level.

Before leaving Vietnam in 1968, Idle received Maintenance Man of the Quarter Award (February-April 1968) for the 8th Air Force.

In July 1968, Idle's tour of duty was over. Soldiers were cautioned against wearing uniforms in airports as war protesters caused problems. The young soldiers wore civilian clothing and didn't encounter difficulties.

In August 1968, Sergeant Idle was discharged from military service at McCord Air Force Base in Washington State. He worked for two years at Warsaw Moving and Storage in Indiana, before a friend who was a member of the Fort Wayne Air National Guard recommended Idle apply there.

After talking to a recruiter, Idle signed up. In 1972, he was assigned as an AGE mechanic and a few months later, he became a technician at full-time status.

Idle worked in the AGE shop on F-100s until 1979 when the base changed to F-4s. In 1991, the base again switched aircraft, this time to F-16s. Idle retired in 1999 at the rank of Master Sergeant E7.

Today, Idle, who lives in South Whitley, Indiana, visits regularly with veterans, especially those he served with.

"Serving in the Air Force during the Vietnam War was a job to me," he said. "I had no thoughts or feelings about the politics. It was an honor to serve. But I can't imagine what my parents went through with me gone for two years. As a parent, I would have a hard time with that."

Master Sergeant E7 Lanny Idle

LeRoy Jesfield -- Army

In the early morning hours of May 10, 1970, LeRoy Jesfield and others of the 221st Signal Company were roused from their sleep and ordered to assemble for a meeting. The meeting was being called on the group's only day off and they weren't happy.

Stumbling in to the mess tent, the sleepy soldiers forgot their disgruntlement at the sight of Captain Bill Kelly's somber expression.

Kelly reported that on the previous day, five members of the 221st and four members of the 189th Assault Helicopter Company had been shot down over Cambodia.

Everyone on board the aircraft had perished.

As the officer shared the names of those who had died, Jesfield was grief-stricken. One member of the 221st had been SP5 Douglas John Itri, a close friend of Jesfield's.

Kelly explained that the explosion had been witnessed by a second American helicopter and upon landing, the crew of the second aircraft had retrieved the bodies of the deceased. Itri's remains would be sent to his family in Boston.

Wanting to do something special to honor his friend, Jesfield created a portrait from a photo he had of Itri. It was Jesfield's first attempt at a portrait created posthumously. He sent it to Itri's family as a tribute to his friend's service.

**

Jesfield was born in 1947 in Billings, Montana. He had enlisted in the Army in 1967 after graduating from Billings Senior High School in Montana.

He was pleased to be assigned to the cryptography (code making) unit as he thought being surrounded by sand bags was a safe place to be in Vietnam.

After training for five months at Fort Gordon in Georgia, Jesfield arrived in Vietnam in 1969. At that time, 500,000 American troops were stationed at various locations in the country, making that time the height of the conflict.

Jesfield was attached to a National Guard unit from Rhode Island. He felt the outfit resembled the misfits from the TV show *MASH*.

He and others in the cryptography unit were given unusual directions. They were told to forget everything they had learned about cryptography in the States. By the time they had arrived in Vietnam, the information was out of date with soldiers in the field advancing from teletype to computers.

Unit members scrambled to learn new material about codes, working 12-hour shifts, six days a week. The cryptography unit had direct access to troop movements and a direct line to American President Richard Nixon.

At first the work in cryptography excited Jesfield. Then, desiring a change, he talked to a friend who was a graphic designer.

While growing up, Jesfield had drawn faces of classmates for enjoyment. He wondered if his skills might be useful as a soldier.

The friend explained that the 221st Signal Company (photographic), Southeast Asia Pictorial Center in Long Binh looked for men with artistic talents to create handmade brochures, slides, maps, and presentations for the news media and films for training purposes.

Jesfield's friend had seen his work and encouraged him to submit drawings. It wasn't long before Jesfield was accepted into the program.

As a member of the 221st, LeRoy Jesfield drew a myriad of pictures, mostly for training of new arrivals. "We provided recruits with materials to help them adapt to Vietnam, what was acceptable to do and say in the country and what was not," he said.

Jesfield also contributed to a comic strip called O.D. (Olive Drab) Strat, depicting a soldier returning to the United States and the situations he found himself in. The comic was printed in the First Signal Brigade newspaper.

Jesfield used pencils to draw his projects, including portraits. He became so adept that eventually he could draw a portrait in an hour.

After 10 months in country, LeRoy Jesfield returned to the United States, but not before receiving an Army commendation medal from General Thomas Matthew Rienzi, First Signal Brigade Commander.

Jesfield was re-assigned to the Third Army Soldier Show based in Fort McPherson in Atlanta. Accompanied by nine other soldiers, he sang, designed sets, and created posters for USO-like shows performed throughout the South.

In November 1971, Jesfield was discharged at the rank of Spec 5 (Buck Sergeant E5). In 1997, he married Wanda Sprunger of Berne, Indiana. The couple attended Berne Church of the Nazarene and LeRoy led singing at a local nursing home.

In 2001, Jesfield resumed his love of drawing, creating paintings and portraits for friends and family, including landscapes, from a studio in his home. When a friend lost a grandson who was killed while serving as a soldier in Afghanistan, Jesfield sketched the soldier's portrait from a photo. "I hoped to honor the family in this way," he said.

Jesfield never forgot his friends in the military. For many years he attended ceremonies on May 9 when Signal Corps veterans gathered to remember the five soldiers who were shot down in 1970.

In 2016, LeRoy Jesfield participated as a veteran with Honor Flight of Northeast Indiana. During the day spent in Washington DC, Jesfield and dozens of other veterans toured the nation's military monuments and memorials at no cost to them.

Jesfield was especially thrilled to be chosen to join a small group of vets in laying a wreath at the Tomb of the Unknown Soldier. "It was one of the most special moments of my life," he said.

On November 27, 2017, LeRoy Jesfield died from colon and rectal cancer.

Buck Sgt E5 LeRoy Jesfield (right) receives an Army commendation medal from General Thomas Matthew Rienzi, First Signal Brigade Commander.

Colonel Dan Lavine -- Marines

In April 1969, Corporal Dan Lavine doubted if he would make it home from Vietnam alive.

The Viet Cong was trying to take the air strip at the city of Da Nang. Having survived the Tet Offensive a few months earlier, Lavine was thankful for the training he had received at boot camp in San Diego on how to shoot M-16s, shotguns, 45-caliber pistols, and 50-caliber machine guns.

When Lavine's left ear was damaged during a battle from artillery and a landmine, he sustained permanent hearing loss.

It seemed a small price to pay in service to his country. Lavine had learned such patriotism from a long line of family who had served in the Marines from cousins and uncles to his future father-in-law.

Lavine, who had left a girlfriend at home in Bluffton, Indiana, just hoped he wouldn't regret his decision to serve his country.

**

He was born in Portland, Indiana in 1948. When his family moved 20 miles north to Bluffton, he graduated from Bluffton High School in 1966.

The following year, Lavine joined 11 other young men from Indiana in enlisting in the United States Marine Corps. He was sent to Marine Corps Recruit Depot San Diego. Thanks to his dedication to athletics in high school, Lavine had few problems with the rigor required by drill instructors. With much hard work and grit he scored a record for push-ups, performing 200 in four minutes.

Among the things that impressed Lavine the most at boot camp was the importance of taking direct orders without questioning. He and other recruits were taught the adage: "When you hesitate, someone pays the price."

The 10 weeks at San Diego was made easier with letters of encouragement from family and friends. When he was overseas, Lavine sent paychecks home.

He completed Individual Training Regiment at Camp Pendleton in California during which troops spent three weeks of survival training in mountains and terrain that resembled Vietnam. On graduation day, Lavine stood on the parade deck with 3,000 other new Marines. He felt gratified when, after ceremonies had concluded, his drill instructor, a staff sergeant, told Lavine he would be glad to have him in his rifle squad.

In April 1968, Lavine arrived in Vietnam where he was assigned to B Company, 7th Engineers, 1st Marine Division. He and other recruits quickly adjusted to living in the hot country where it seemed no one could be trusted.

Lavine observed that the Viet Cong fought in surges, rather than daily, probably due to a lack of supplies. In between battles the Americans built roads and bridges from the DMZ to Saigon, blew up tunnels, and did mine sweeps. The sweeps were challenging, as roads weren't paved, just packed dirt.

They worked and fought along Hill 34 and Hill 55, west of Highway 1 on the outskirts of Da Nang.

Weather presented its own challenges, especially during the rainy season when troops slept on pallets to avoid getting wet.

Meals were nothing to write home about -- often they were sea rations leftover from World War II two decades earlier.

Lavine survived the battle at Da Nang and was told he could leave Vietnam in May 1969. He departed, saddened by the knowledge that only five of his original unit of 12 enlistees from Indiana were still alive.

Although Lavine was anxious to get home after being gone for 13 months, he and other soldiers were taken to Okinawa for a week to have their emotional stability evaluated. When all were deemed well, they were allowed to return to their homes.

Lavine had liked the Marines enough to remain in until 1973 when he joined the Army National Guard unit in Bluffton, Indiana.

Over the next two decades, Lavine attended Officer Candidate School at Indiana Military Academy at Atterbury and participated in deployments to Germany, England, and Czechoslovakia. By the time he retired in 1999, he had reached the rank of full colonel.

He also worked at Franklin Electric in Bluffton where he used leadership skills obtained in the military for tasks.

Today, Lavine has a great feeling of national pride upon hearing 'The Star-Spangled Banner'. "Hearing that song causes the hairs on the back of my neck to raise," he said. He also believes Americans should be grateful to live in this country. "Most Islamic countries envy Americans because they think we are rich," he added. "But it's only because God has blessed us. We soldiers were fighting to help people have a better way of life."

Colonel Dan Lavine

Rodney Maller – Army

Rodney 'Rod' Maller stood, his medic aid bag packed for a combat patrol.

He had been assigned to HQ 1st Infantry, 2nd Battalion, 18th Infantry, Company C. Upon arriving in Vietnam in June 1969, Maller, as the unit's corpsman, was responsible for being prepared for any eventuality in caring for possible injuries for the men in his unit.

He was admittedly nervous, despite receiving months of training. But he wanted to prove his willingness to support his men.

Thus, it was a blow when Maller's sergeant replaced him with another medic. "You're not ready," he told Maller.

Maller, 21, didn't argue, although he felt disappointed at being left behind. Then he heard what happened to the unit.

The Viet Cong ambushed them. The medic who had taken Maller's place was killed. "It was a tough start to my tour," he said.

**

Maller was born in Adams County in 1948. After graduating from Adams Central High School in 1966, he worked as a mechanic at a local car company and attended mechanic school in St. Louis, Missouri, for a year.

Upon being drafted into the Army in February 1969, Maller registered as a conscientious objector for religious reasons (he attended an Apostolic Christian Church). He was assigned as a medic, despite having no previous experience in the healthcare field. "We were told medics were one of the Army's biggest needs," he said.

Maller received training at Brooke Army Medical Center (then Fort Sam Houston Medical Training Center) in San Antonio, Texas. The medical facility was the Army's largest in the country and the only military base that provided training for medics headed to Vietnam.

For two months Maller and other medics-in-training learned how to administer first aid, CPR, and bandages. They applied splints and sutures, IV's, shots, and learned how to handle everything from accidental knife cuts, heat stroke, spider bites to punji stick injuries. Every GI carried thick bandages for the most basic injuries, but the cry of 'Medic!' could be heard often, especially on patrols.

Soon, Maller was permitted to accompany a patrol and he was placed in the middle of the unit's single-file march formation. "It was thought the front and back were the most dangerous positions if we were caught in an ambush," he said. Every soldier knew his welfare depended on the medic. As a conscientious objector, Maller didn't carry a weapon. He felt confident in relying on the GI's around him for protection.

Possible types of injuries varied greatly. During one patrol, a soldier peered into what appeared to be an old bag on the ground before Maller could warn him to steer clear. A heavy vapor filled the air, causing the soldier's eyes to tear. He clutched at them, hollering in pain.

Maller quickly instructed a sergeant to hold the soldier's now-reddened eyes open while he poured water from his canteen to flush them. Maller assessed the bag contained a noxious agent, possibly tear gas tossed from a plane.

Maller continued the water treatment until the soldier could see, though his vision was fuzzy.

The soldier was evacuated via helicopter and Maller later learned the doctor on the chopper approved of Maller's immediate cleansing of the eyes. Without it, he believed the soldier would have been dead within two hours as the gas would have entered his bloodstream through the eyes.

Another time, when a soldier's eye was punctured by a thorn, Maller recalled from training that the plant could have a hook. It should not be yanked, as the action could rip the eyeball.

Maller bandaged the eye and sent the soldier to a hospital where a surgeon carefully removed the barb without damaging the eyeball.

Hot weather caused special health concerns for the troops. Chopper pilots dropped food rations and cylinders of water to fill soldiers' canteens. When the water supply ran out, Maller tested streams running from mountains for bacteria with chlorine tablets. He also encouraged troops to wear wet towels around their necks to stay cool.

Soldiers stricken with heat stroke were laid on the ground and covered with ice also delivered by chopper pilots.

Once, when an officer stopped breathing due to heat stroke, Maller ordered that he be sent back for medical attention. The officer begged Maller not to send him off the field.

Ground troops use air transport for patrols. Courtesy Rod Maller.

The officer's dedication to his men and duties impressed Maller, but he believed the severity of the officer's condition warranted medical attention and he carried through.

Maller witnessed that same level of commitment to duty in most of the soldiers he met. "It was uncanny how dedicated they were to our cause," he said. Soldiers within one month of going home were a different story. "Then they often got a little crazy," he added.

While on patrols, troops had trouble keeping their feet dry and avoiding foot rot. Pilots dropped foot powder and clean socks, but Maller chose another method for his own self-care to prevent problems. He didn't remove his shoes for seven months.

"I thought my feet would stay safe from foot rot if I kept them on," he said.

Troops slept outdoors, sometimes in rice paddies to hide from the enemy. During the monsoon season from October through March, rain pounded hard enough to sting the skin and rice paddies filled with water. The troops often awoke, faces covered, scrambling from paddies so as not to drown.

During patrols of villages, troops searched huts for signs of the enemy. "We could tell from the looks on villagers' faces if they were harboring the enemy," he said. "They looked threatened and frightened." Some huts had dugout bunkers inside for the family to be out of the line of fire during a fight.

Maller and other American medics were authorized to treat villagers who usually appeared thankful. "Even giving a bandage or an aspirin for a headache was a big deal to them," he said.

Maller's most heart-wrenching experience occurred when a mother approached him holding her child. Maller knew from the woman's motions that she wanted him to heal the infant who he suspected from the little one's sickly countenance would not live long. Knowing it was too late, he grieved at having to turn her away.

At the end of two-week patrols, troops returned to their fire support base for a day of rest and the opportunity to sleep inside of a building, eat a hot meal, and take a shower.

At the base they slept on pallets to keep away from rodents. However, one night, Maller fell asleep on a pallet, only to awake and find a rat on his chest. He shooed it away, relieved it had not bitten him.

Another night, Maller was nearly injured when the Americans were preparing for an ambush. He moved up behind the point man to be ready when a piece of shrapnel (presumably from the enemy) took a chunk of plastic from his glasses. He felt dizzy but was otherwise unhurt.

After seven months of patrols, Maller was transferred to the 25th Infantry Division operating next to the Cambodian border near Tay Ninh. His duties consisted of driving an ambulance.

The 25th's location was in a section called "Angel Wing" close to Saigon. Although not harassed by enemy troops, members of the 25th were exposed to another danger.

The American government ordered that the area be sprayed liberally with an herbicide. It was an effort to protect residents and troops guarding the city by killing the tall grass surrounding it, thus enabling troops to see signs of the enemy.

Maller thought little about the liquid nicknamed Agent Orange that came down in a mist until decades later. Then it would have a major impact on his life.

Upon receiving an early release from service in June 1970, Maller was discharged from the Army in November 1970. He endured flashbacks about the war for a year, but they eventually ceased.

He felt good about the work he had performed as a medic. As much as he was aware, he had never lost a GI. He would have liked to have known the condition of his patients after he put them on choppers. But Maller had to be satisfied in knowing he had tried his best.

Back in Adams County, Rod Maller worked at a farm implement dealership before starting a business, Craigville Diesel Service, with family members in 1975.

In 1978, Maller married and he and his wife, Pat, became parents to a daughter.

Today, Rod Maller is retired. He battles a number of health problems, including diabetes, kidney, heart, and lungs. He believes his experiences in Vietnam were good. "I learned how to listen and have respect for authority and work as part of a team," he said. "We learned over there that what you learned in AIT was the basics. You had to think on your feet and follow orders. Everyone should go through military training because it would benefit them."

Corpsman Rod Maller

Tom Paxson -- Army

The Vietnamese barber spoke fluent English. Chatting with his American customers, he often relayed how he had spent time in the United States prior to the war.

Tom Paxson of Bluffton, Indiana, liked the barber whom American troops flocked to for haircuts. But the friendly stylist had a secret.

One night, Americans troops received small arms fire from enemy forces in a cemetery across the road from their base. Paxson and others leveled the area.

The next day the barber's body was found among the rubble. He had been a member of the Viet Cong. "We learned never to trust anyone," said Paxson. "Charlie could be listening."

**

Tom Paxson had enlisted in the Army in the summer of 1966. He completed basic training at Fort Knox in Kentucky where the recruits were not allowed to call their rifles "guns", but "weapons."

At Fort Leonard Wood in Missouri, tests for Paxson's Advanced Individual Training (AIT) showed he had a mechanical aptitude. His military occupational specialty (MOS) became that of grader, or heavy equipment operator. Tom's father, Fred Paxson, had performed similar tasks as a Seabee in the Pacific during World War II.

Upon arriving in Vietnam in December 1966, Paxson was assigned to the 46th Engineers at Long Binh post near Saigon. The 46th used D7 Caterpillar bulldozers (the crew referred to the machines as 'cats') and Clark 290M earth movers to clear jungles, as well as build roads, runways and foundations for buildings. The unit also used semi-trailers and dump trucks to build an ammunition dump.

Such projects were a strain for the crew, as underneath three feet of solid soil was muck like Jell-O. Paxson offered advice to new drivers to helped them avoid getting stuck. Monsoon rains created their own devilry.

As if environmental challenges were not enough, Paxson's unit was often fired on by the Viet Cong. Troops working on roads knew they were easy targets, but they believed the enemy wanted what they were creating and would be patient until the war was over.

The Americans carried M-14 rifles, though it was not easy returning fire from inside the bulldozers' cabs.

The enemy used M-16s, which were light and carried more ammunition than the M-14s, but Paxson believed the latter had more fire power and dependability.

Another formidable force threatened troops around the base: baboons. "We didn't see many, but enough to keep us on our guard," Paxson said.

In January 1968, Paxson traveled home for a 30-day leave. He surprised his mother, Audra, by walking into her classroom at Columbian Elementary School in Bluffton where she taught third grade.

She was delighted to see him at home – until Tom stated that he was only home for 30 days. He had extended his time in Vietnam for six months. Audra Paxson was upset and tried to talk him out of it, but Tom said he knew what he was doing and had friends there.

He reasoned that as he was not working in the jungle on patrols, he was fairly safe. He made it a point not to re-enlist as he didn't want to go to Germany.

Paxson spent six months in Vietnam, adjusting again to bugs floating in the drinking water (he chose iced tea so they were not as noticeable). At one point, he attempted to bring a bit of his Midwestern roots to the Asian country by growing a garden on the roof of his bunker. The project was short-lived when the military sprayed a substance over the compound, killing his sprouts.

The purpose of the spray was to kill the tall elephant grass so American infantrymen could see the enemy.

Unfortunately, the spray, which became known as Agent Orange, was later blamed for causing a variety of health problems for the troops and people in the area.

Sgt E5 Tom Paxson went home for another 30 days in August 1968, returning in September for his final tour. He left Vietnam for good on March 15, 1969, which was his birthday. "It was the best birthday present I ever had," he said. He served two years and seven months of a three-year enlistment.

Paxson married and he and his wife, Cindy, became parents to two children. He worked 33 years for the Bluffton Fire Department and drove an ambulance for the Wells Community Hospital in Bluffton.

In recent years, Paxson has experienced various health maladies, including neuropathy, diabetes, skin cancer, and atrial fibrillation. But he is proud of his time as a soldier in Vietnam. "It was a war we should have never been in, but we have a great country and I'm proud to have served," he said.

Sgt E5 Tom Paxson

John Senac, Jr. — Army

In summer 1970, John Senac and his trained scout dog, Diablo, quietly crept along a brushy path at Fat City near Vietnam's east coast. A patrol of 10 American soldiers followed closely scouring the area.

Suddenly the German Shepherd stopped, ears cocked forward. Slowly, his back end lowered and his head turned slightly to the right. Senac understood his dog's signals.

Taking a knee, he waved the squad leader forward. Senac quietly informed the officer that Diablo had caught the scent of enemy soldiers ahead. He recommended the squad move perpendicular at nine o'clock for 100 yards, then prepare to engage with personnel at two o'clock. (Times on a clock were used to signify direction)

The officer shook his head. "We're going straight ahead to the hill," he whispered back.

Senac tried to convince the second lieutenant of the danger ahead, but he refused to listen.

Senac had been taught during dog handler training that if a squad leader negated the alert of a handler and scout dog, the handler could come off point and request a helicopter for safe extraction of him and his dog.

Believing the patrol was headed into an ambush, Senac stated that he and Diablo were going off point and wanted a chopper to return to base. The officer agreed, albeit reluctantly.

Senac pulled Diablo aside, whispering to each soldier that passed: "Personnel at two o'clock. One hundred twenty five meters. Hedges and shrubs."

The soldiers looked confused but proceeded. Seconds later, enemy rounds arrived, waist-high and close.

The point man's screamed filled the air. Senac shuddered. It was the first time he had seen and heard an American soldier get wounded in battle.

Unable to watch more men get slaughtered, Senac shouted, "Watch my tracers!" He fired his CAR-15 three times in quick succession to mark the enemy's location, then pulled Diablo back as the squad sent bullets towards the area where the Viet Cong hid.

The patrol's medic had pulled the wounded soldier back and Senac heard him mutter that a 7.62-millimeter bullet had hit a bone in the soldier's right leg before exiting. The medic administered a shot of morphine and applied a tourniquet to the limb while consoling the pain-filled soldier. "Hey, pal, you're going home! You got a million-dollar wound!"

The soldier was loaded on a medivac and by the time the bird lifted, the battle had ceased with an enemy retreat.

Senac and Diablo arrived safely at the base camp, thankful no other members of Senac's squad had sustained injury. But Senac seethed at the squad leader's disregard for Diablo's alerts. Senac knew he and Diablo as a team could help to save men's lives. But that was true only if others respected and honored their training and contributions.

**

John Senac, Jr. had been born in New Orleans, the oldest of five children. After being drafted in March 1969, Senac completed boot camp and Advanced Individual Training at Fort Ord, California. At Fort Benning in Georgia Senac began training as a non-commissioned officer (NCO).

That changed when Senac heard about the need for dog handlers. His family had owned dogs and Senac felt comfortable around canines. He quit NCO training to pursue the duty of scout dog handling.

The effective work of scout dog handlers and their canines caused the enemy to place bounties on their heads. Courtesy John Senac, Jr.

During 12 weeks of training, Senac and the dog he was assigned to learned how to detect airborne body scents when American soldiers dressed as the enemy hid in tree lines during patrols. Other dogs were taught to detect booby traps and mines in the ground. German Shepherds were the most common breed for scout dogs with Great Danes, Collies, and mixed breeds also used.

Their schooling took place in the undeveloped, bushy areas of Georgia and Alabama where the hot, humid, lush climate provided a substitute for the triple-canopy effect of Vietnam. On weekends the teams returned to base for hot meals, showers, and rest.

On January 6, 1970, after completing his training, Senac left his scout dog (each dog worked with three handlers) and flew to Tan Son Nhut Air Base. There, he was matched with Diablo at the 57th Scout Dog Platoon Americal Division.

Senac was one of 10,000 soldiers eventually trained as dog handlers and Diablo one of approximately 4,244 scout dogs.

As Senac and Diablo spent much time together, Senac grew to understand Diablo's alerts. The canine could detect the enemy's scent at 100 yards.

With keen skills and training, scout dogs and their handlers became threats to the enemy. When intel revealed that bounties had been placed on their heads, the 57th provided a rear area with kennels.

Enemy fire was not the only danger to the scout dogs.

During the summer of 1970, Diablo contracted a disease called Idiopathic Hemorrhagic Syndrome. More commonly called "Red Tongue", it involved inflammation of the stomach and intestines with bleeding and vomiting, often resulting in serious consequences.

The illness, so called because it changed the color of a canine's tongue, became a common malady to American military dogs in Vietnam. After receiving a full blood transfusion, Diablo appeared well enough to return to duty. Unfortunately, his health again declined, resulting in his death in October 1970.

After the loss of Diablo, Senac felt stricken but had little time to grieve as he was immediately paired with another dog. He helped with the training of new dog handlers, now arriving at a rate of 100 per week.

In fall 1969, the American government began sending soldiers home short of their full deployments. When E4-Spec Senac's original date was moved from January 1971 to December 4, 1970, he arrived in New Orleans in time for Christmas. He began working in the construction industry, a career that would carry him through to retirement.

Scout dogs in Vietnam were also reduced, but not by returning to the United States. Many canines were given to the South Vietnamese military. Those considered "unessential equipment" were euthanized. Approximately 350 scout dogs died either by Red Tongue or combat.

Senac married and he and his wife became parents to five children. They lived for many years in Indianapolis.

Today, Senac and his wife live near family in Bluffton, Indiana. In 2021, Senac became the proud owner of another German Shepherd, Thor.

Despite the losses of friends and his beloved Diablo, Senac was glad to have served in Vietnam. "It was amazing to know each time we were on point we found something before Charlie found us," he said. "We never lost a man or animal."

E4-Special John Senac, Jr. and Diablo

Lt. Colonel Harold Stanford

Army

Two American helicopter pilots circled the mountains east of Vung Tau in southern Vietnam, speaking on the radio with the operations sergeant as though preparing for another assault. Then one flew south of a mountain while the other headed west. The movements caused a brief pause in communication.

During that split second, a male Vietnamese voice began chattering noisily, non-stop and nonsensical.

It was just what Harold Stanford, operations officer to the 25th Vietnamese Infantry Division, had hoped would happen.

For the past several months American pilots had been interrupted by an unidentified male voice while trying to communicate during crucial operations. The stranger (no one knew who he was) took over the airwaves, interfering to the extent that aircraft had been badly shot up.

Harold Stanford's commander was frustrated. It was vital that soldiers prepare for a combat assault using radio communication. A trusted interpreter to the Americans said the interloper was reporting to someone on the radio – presumably the enemy --

about the missions, including number of American helicopters, directions they were headed and more.

Stanford's duties at the 25th included organizing helicopters for combat assaults and coordinating with the Air Force to make airstrikes on landing zones. No one knew the location of the interfering man.

Stanford nonetheless approached his commander. "Give me two slicks (helicopters) and a gun team and we'll get rid of the problem," he promised. Stanford's boss had looked skeptical but did as requested.

Stanford had instructed the pilots of two helicopters to use their FM homing devices to get a fix on the interloper's location, while appearing to prepare for a mission. The homing devices showed the stranger was on the top of a nearby mountain.

Stanford then instructed the pilots to go up the following day and get the stranger talking while 'heavy hogs' (helicopters with rocket pods that held approximately 24 rockets each) opened fire.

The plan proceeded without a hitch. Afterward, Stanford's crew was satisfied at finding nothing but an antenna hanging in a nearby tree.

**

Harold Stanford grew up in Lineville, Alabama. After graduating from Jacksonville State University in Alabama in 1959, he received his commission into the Army, entering active duty at Fort Bliss in Texas.

In 1960, Stanford was assigned as Launcher Platoon Leader to Site W92, a Hercules Missile Battery in Rockville, Maryland. That same year he attended flight training and was assigned as a helicopter pilot at Pershing Missile Battalion at Fort Sill, Oklahoma.

For six months in 1964, Stanford was assigned to a helicopter unit at Fort Benning, Georgia. Finally, with years of extensive training, he deployed to Vietnam.

At Duc Hoa in the Mekong Delta in southwest Vietnam, the army air field was constantly busy with air traffic. Everyone was cautioned to stay alert for signs of the enemy.

One day, Stanford stood on the runway performing a pre-flight check on a helicopter when he noticed a chopper coming in, nose up. Not far away he noticed another chopper taking off. Its nose was low and Stanford could tell from the speed and direction of the two aircraft that the pilots were unaware of each other.

He watched in frozen horror as the two collided, bursting into flames. Stanford and others in the area raced to pull the pilots from the burning chaos.

When Stanford knelt to pick up a helmet, he gasped at the sight of the head inside. He turned away, retching and sorrowful for the pilots who had perished.

Dreadful memories from the fatal crash stuck with Stanford and he didn't eat for days. Finally, he consulted the base's flight surgeon. Maybe he could suggest a treatment.

The doctor listened to the pilot's dilemma. Then he set a bottle of whiskey and two glasses on his desk. "Have a drink," he said.

"Doc, I'm not thirsty. I'm hungry."

The doctor poured the liquid. "Maybe if we both drink enough of this, we can eat. I helped clean that mess up, too."

Stanford felt reassured that he was not the only one traumatized by the gruesome event.

He and the other Americans learned that the Viet Cong was even more lethal, especially at rigging bombs. They were warned to watch for anything out of the ordinary.

Flight crews search for the enemy on search and destroy missions. Courtesy Rod Maller.

When Stanford found a bicycle propped against a wall near his bed, he called military police. They, in turn, notified the explosive ordinance unit who found explosives packed in the bike's frame. They carefully removed it.

Another night, while talking to a soldier in the hooch (barracks), Stanford pulled out a box of grenades he kept under his bed in case of attack. Somewhat distractedly, he unscrewed the fuse of one, looked at it, poked it back into the grenade, and screwed it back in. Note: A grenade is 'live' when its pin is pulled.

Still chatting, Stanford repeated his movements on another grenade from the same box. He paused before replacing it. It looked slightly different.

Stanford removed fuses from all of the grenades in the box, laying them on the bed. All were from the same lot number and should have looked exactly alike. Instead, several had shorter fuses.

He knew the shorter fuses, if thrown, would explode immediately, causing damage to a hand, ears or a concussion – maybe death.

Stanford was convinced the manufacturer had not made such a careless mistake. The only people with access to the soldiers' hooches were soldiers who lived there and Vietnamese maids who washed clothes and cleaned the floors.

One day, pretending to sleep in his bunk, Stanford noticed a maid enter, pull the box from under his bed, and switch the fuse on a grenade.

After she left, he reported her actions to the division commander. Stanford was told the matter would be handled and to stay vigilant.

Three days later, he saw American military guards take the maid to the commander's office. A few minutes later, he heard a gunshot. Stanford watched as guards dragged the woman's limp body behind a building, presumably to bury it.

In January 1966, Stanford returned to the United States to attend Artillery Advanced Course at Fort Sill, Oklahoma, and Fort Bliss, Texas. From

October 1966 through September 1968, he was assigned to a CH-47 (Chinook) Unit at Fort Benning, Georgia, before being deployed again to Vietnam.

During one mission in a CH-47 Chinook, Stanford hauled 8,000 pounds of ammunition to an American unit in contact with the Viet Cong. Upon calling to ask in which direction to land, Stanford was told east to west.

Coming in at that position, he received ground fire. Forced to drop his load, he departed south while attempting to fly though enemy fire which disabled one engine and his hydraulic and electrical systems.

Stanford, knowing he no longer had a workable aircraft, issued a *Mayday* call on the radio. His tension heightened when his co-pilot and two other members of the crew sustained injuries.

Somehow Stanford kept the aircraft airborne for 10 miles and was greatly relieved when gunship support arrived from Cobras, enabling him to land safely.

For their actions that day the wounded crew members, all of whom recovered, received Purple Hearts.

Harold Stanford received the Distinguished Flying Cross for 'heroism while participating in aerial flight evidenced by voluntary actions above and beyond the call of duty.'

Not every military action was treacherous.

Before leaving the States, Harold Stanton obtained several 100-kilowatt generators from an Air Force supply center. Upon arriving in Vietnam, he gave one to each of his pilots to cool their hooches. He kept the last one for a special purpose.

When his unit was inactive, Stanford flew the unit's flight surgeon to a leper colony. Many pilots were afraid of being exposed to the skin disease, but several helped to deliver caches of rice to the group.

Within a year, the surgeon had arrested every case of leprosy in the colony. As an extra benefit, Stanford placed the extra generator in the colony and a light bulb in each house. It was the first time the villagers had access to electricity.

In 1968, after two tours in Vietnam, Stanford returned to the United States. During the next two decades, he was assigned to various positions, including Commander of Godman Army Airfield at Fort Knox, Kentucky and Aviation Officer and Commander of Biggs Army Airfield at Fort Bliss, the largest Army airfield in the nation.

In 1978 he was deployed to Kuwait for one year as Air Defense Advisor to the Kuwait Air Force.

During his absences, Stanford's wife, Martha, raised their two sons, David and Mike Stanford.

Lieutenant Colonel Harold Stanford retired from the military in 1988. He worked as a corporate pilot in Louisville, Kentucky for several years. Today, he resides in Barfield, Alabama, where he is a member of the volunteer fire department and he and Martha attend Mannings Chapel.

While Harold Stanford's contributions to his nation are not known by every American, he is content being thought of as an "unsung hero."

Lt. Colonel Harold Stanford

Glossary

Agent Orange: Herbicide mixture used by the U.S. military during the Vietnam War containing a dangerous chemical contaminant called dioxin shown to have harmful effects on people exposed to it. The substance may have received its nickname due to the orange stripe on the drum in which it was stored. Production of Agent Orange ended in the 1970s and it is no longer in use.

Advanced Individual Training (AIT): Skilled training which takes place after a military service member completes basic training, referring to instruction received in an assigned military career field.

Army of the Republic of Vietnam (ARVN): Ground forces of the South Vietnamese military from its inception in 1955 to the Fall of Saigon in April 1975.

Charlie: Abbreviated term for the phonetic spelling used by the military to spell names over the radio, as in "Victor Charlie" for "Viet Cong", eventually shortened to Charlie.

demilitarized zone (DMZ): Nearly 50-mile border between North and South Vietnam running east to west in the middle of present-day Vietnam, approximately five miles wide.

fire base: Military camp equipped with big guns.

Ho Chi Minh City: Name given to Saigon, the capital of South Vietnam, in 1976 after the country was reunified to honor the North Vietnamese leader Ho Chi Minh.

punji stick: Wooden stake or spike sharpened and camouflaged by grass for purposes of decimating one fighting group by another. Its tip may be poisoned or covered with dung.

Tet Offensive: A coordinated series of North Vietnamese attacks on more than 100 cities and outposts in South Vietnam from January 1968 through September 1968.

triple-canopy: Dense ceiling of leaves and branches formed by closely spaced forests.

Viet Cong: Members of the communist guerrilla movement in Vietnam that fought the South Vietnamese government forces 1954–75 with the support of the North Vietnamese army and opposed the South Vietnamese and U.S. forces in the war.

Book Discussion Questions

1. Mike Boles was chosen by his unit to be the "tunnel rat." What thoughts might have gone through his mind at being chosen for such a dangerous job?
2. Tom Paxson didn't want to return to the U.S., due to war protesters. How did soldiers who did return get treated? Give personal examples if you lived through this period.
3. Harold Stanford provided a generator for a leper colony in Vietnam. How did other veterans in the book respond to the needs of Vietnamese citizens?
4. After falling into a crater following his head injury, Rick Dawson lay stunned with no vision or hearing until a medic found him. What other examples are in the book of soldiers helping each other?
5. Jose Huerta's family provided military service for many generations. Name another family you're familiar with who has done the same.
6. Mike Chamness struggled with post-traumatic stress syndrome upon returning to the U.S. He later addressed the problem by writing poetry and attending group

counseling for anger issues. What other options are available to veterans today?
7. Many Vietnam War veterans were not required to serve a year of service. This was different from veterans in earlier conflicts who served longer periods. What advantages or disadvantages would there have been with each group?
8. Which story resonated with you? Explain.
9. Tell a story about a veteran you have known from Vietnam or any war.
10. What act of service might you perform to show appreciation for military veterans?

Excerpt from Mary Anna Martin 'Marty' Wyall's story in *Born To Be Soldiers: Those Plucky Women of World War II*:

In 1939, the Martin family was living in Seymour, Indiana, where Mary Anna graduated from Shields High School. She spent a year at MacMurray's Women's College in Jacksonville, Illinois, on a music scholarship before transferring to DePauw University near Greencastle, Indiana, to major in bacteriology.

During her sophomore year, Mary Anna heard about the terrible events at Pearl Harbor on December 7, 1941. She was moved to tears upon hearing radio reports of the 2,403 men who were killed and hundreds more injured.

Mary Anna vowed to help her country in a meaningful way, she just didn't know what it would be.

Then Mary Anna read a magazine article about a new program with the Army – one that was recruiting women to learn to fly military aircraft.

Mary Anna was intrigued with the idea. The women would be called 'WASP' – Women's Air Service

Pilots. Their training would allow more men to go to war.

Mary Anna wanted to apply to the WASP program. Her father insisted that she graduate from college first. Mary Anna obeyed – mainly because she needed time to accumulate the 35 flight hours required of applicants.

After graduating from DePauw, Mary Anna worked at Eli Lilly, a pharmaceutical firm. She saved money from her earnings for flying lessons which she took at Sky Harbor in Indianapolis.

When Mary Anna had logged 35 hours of flight time, she completed the WASP paperwork and submitted it. She was thrilled to be informed she had been chosen from among the 25,000+ other applicants to become a member of the WASP. Her flight instructor was happy for her as well.

Not everyone wanted Mary Anna to join the WASP. Her mother, Bernice Martin, thought it was wrong for a woman to be in the military. She refused to say good-bye to her daughter on the day of Mary Anna's departure.

Books by Kayleen Reusser

World War II Legacies:

We Fought to Win: American WWII Veterans Share Their Stories (Book 1)

They Did It for Honor: Stories of American WWII Veterans (Book 2)

We Gave Our Best: American WWII Veterans Tell Their Stories (Book 3)

We Defended Freedom: Adventures of WWII Veterans (Book 4)

World War II Insider:

D-Day: Soldiers, Sailors and Airmen Tell about Normandy (Book 1) and *Battle of the Bulge: Stories From Those Who Fought and Survived* (Book 2)

Captured! Stories of American WWII Prisoners of War (Book 1, Prisoners of War)

It Was Our War Too: Youth in the Shadows of WWII (Book 1, Witnesses of War)

Women of WWII Coloring Book

Men of WWII Coloring Book

About the Author

Kayleen Reusser has interviewed hundreds of veterans, including 260 from World War II. She has written several books and completed research in Europe. As a speaker (virtual and in-person), Reusser has presented talks about American military topics across the nation.

Subscribe to her Youtube Channel to hear short talks by veterans she has interviewed.

Her books are available on Amazon.

Go to her website to sign up for her FREE email newsletter to hear about her latest projects.

www.KayleenReusser.com.

www.ingramcontent.com/pod-product-compliance
Lightning Source LLC
Chambersburg PA
CBHW060322050426
42449CB00011B/2612